A new key to wild flowers

A new key to
WILD FLOWERS

JOHN HAYWARD

with illustrations by Michael Hickey

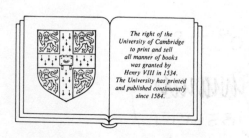

The right of the
University of Cambridge
to print and sell
all manner of books
was granted by
Henry VIII in 1534.
The University has printed
and published continuously
since 1584.

CAMBRIDGE UNIVERSITY PRESS
Cambridge
London   New York   New Rochelle
Melbourne   Sydney

Published by the Press Syndicate of the University of Cambridge
The Pitt Building, Trumpington Street, Cambridge CB2 1RP
32 East 57th Street, New York, NY 10022, USA
10 Stamford Road, Oakleigh, Melbourne 3166, Australia

First published 1987

Printed in Great Britain at the University Press, Cambridge

British Library cataloguing in publication data
Hayward, John
A new key to wild flowers.
1. Angiosperms – Dictionaries
I. Title
582.13'03'21   QK496.Al

Library of Congress cataloguing in publication data
Hayward, John, 1922–
A new key to wild flowers.
Includes indexes.
1. Wild flowers – Identification. I. Title.
QK85.5.H39 1987   582.13   86–4249

ISBN 0 521 24268 1 hard covers
ISBN 0 521 28566 6 paperback

WD

CONTENTS

PREFACE

These keys were started when I used to run Field Centre Courses on Wild Flowers for amateurs. I wanted to introduce keys to those who were keen to name their finds but had never got beyond the stage of looking at illustrations (preferably coloured ones). The full keys of those days, Bentham and Hooker, and later 'C.T. & W.', were much too technical and long, and entirely unsuitable for carrying in the pocket. So I would introduce the Course by handing out to each student a dozen different species of 'yellow-Dandelion-like-flowers' and a key (The Simplified one on page 260) in order to establish that once the method was understood and a few terms explained any of the very similar-looking flowers could be identified. The other Simplified Keys were produced at daily intervals. They were much in demand and even brought back by students returning the following year . . . and so, like Topsy, the book 'just growed'.

I should like to acknowledge the patience and forbearance that I have received from the publishers, the willingness to draw and redraw that the artist, Michael Hickey, has shown, and the practical help that I have been given by the AIDGAP Organization, particularly from Dr Anne Bebbington who has used, perused and tested the keys at Field Study Centres and made suggestions aimed at improving the convenience, accuracy and 'user friendliness' of the whole work.

It must not be expected that keys set out like these, with a minimum of words and illustrations, will provide an answer on every occasion – plants are too variable. But it should be only very rarely that the correct name cannot be found.

Read the pages 'How to use the key' carefully and then the 'General notes' before beginning.

THE AIDGAP PROJECT

Putting a name to an animal or plant remains a fundamental part of most forms of biological fieldwork. It can also be the most difficult and frustrating aspect of such work, particularly when dealing with a group of animals or plants for the first time. The AIDGAP project is designed to help biologists with groups in which the difficulty is due to the absence of a simple and easy-to-use key.

The Field Studies Council, which administers the AIDGAP scheme and runs nine residential Centres, all of which specialise in fieldwork, is in a unique position to identify those groups for which the difficulty in identification is due to the absence of a suitable key. A unique feature of AIDGAP keys is the way in which they are processed before final publication. All are 'tested' extensively in addition to the routine editing and refereeing by acknowledged experts. Several hundred copies of a preliminary draft – the 'test' version – are distributed to a wide range of users. This group will include both academics and amateurs, botanists and zoologists, teachers and students alike, and their comments after a year's testing are compiled to form a report which is sent to the author who then revises the final publication. This process ensures that maximum ease-of-use is combined with scientific accuracy.

Publication of this new key to wild flowers is a milestone in the development of the AIDGAP project. It results from the first formal co-operation between AIDGAP and a major publisher. Numerous copies of an earlier draft were subjected to extensive field trials, largely at Field Studies Council Centres, and this version has been revised accordingly. We hope that the key now meets the requirements of fieldworkers but Cambridge University Press would be pleased to hear from those with suggestions for possible improvements. The success, or otherwise, of keys such as this depends largely on feedback from the people who use these publications.

All of the AIDGAP project's previous titles have been published under the Field Studies Council's imprint. A complete list, and further details of the AIDGAP project and FSC Courses, can be obtained from: AIDGAP, The Leonard Wills Field Centre, Nettlecombe Court, Williton, Taunton, Somerset TA4 4HT, UK.

HOW TO USE THE KEY

The principle on which the key works is as follows:

To identify a plant start at the top of the left-hand column and read downwards until the first true statement is reached (true that is of the plant in question); keeping opposite the frame in which this statement appears move into the second column and read down the statements there; if necessary repeat the process in the third column. The name required will then be found in the last column.

Example

The steps in identifying a Primrose, were it necessary, would be as follows:

Turn to the FIRST KEY on page 14.

Under the heading FLOWERING PLANTS the first statement 'Plants w. woody stems' is not true; the next statement 'Aquatic plants' is not true either; nor is the next; and so on, until the statement 'Fls. w. distinct sepals & petals' is reached. This is true so move across to column 3.

The first statement there 'petals joined to each other at least at base' is true because the five petal lobes all spring from a common tube.

Turn now, as indicated in the last column, to PART N on page 47.

Check the heading of this page and then work down column one as before. A quick glance at each statement in turn will show that the first true one is 'Lvs. radical only'.

Move into column 2. A count of the stamens in one flower will show that there are five.

Move into column 3. A count of the styles (one only) will give the family name PRIMULACEAE page 150.

Check the information under the family name on this page and proceed as before.

There are only two statements to be read in the first column and two in the second before reaching the name of the plant *Primula vulgaris*.

2   Of course if it was known that the plant being examined belonged to the PRIMULACEAE a start could have been made straight away by using the INDEX TO FAMILIES at the very end of the book and then turning to page 150.

Remember

1 The plant diagrams on pages 6 and 7, the glossary on pages 8–12, and the list of abbreviations on page 13 explain how botanical terms are used in this book.

2 The statements in each column *must be read in strict order from the top*. The key will not work otherwise.

3 Some large families have been split into sections. The key to sections *must also be taken in order*.

4 In cases of doubt it will not usually matter whether any statement is regarded as true or not. The key will work whichever conclusion is made. Allowance has been made for normal variation in such things as colour, size, hairiness, etc. Some plants will therefore appear in two or even three different places in the name column.

5 If the family to which a plant belongs is known there is no need to work through the whole key. A start can be made on the appropriate family page by using the INDEX TO FAMILIES at the very end of the book.

6 The index (whether family, genus or English name) always indicates the page on which the appropriate key begins.

GENERAL NOTES

Aggregate species Occasionally a plant in the key has the abbreviation (agg.) after the Latin name. This means that the name either covers an aggregate of two or more species, scarcely distinguishable from each other, or else includes a number of microspecies grouped under one name.

The following aggregate species in particular include a fairly large number of microspecies: *Alchemilla vulgaris*, *Euphrasia officinalis*, *Hieracium murorum*, *Rubus fruticosus*.

Aquatic plants Because many aquatics are so often found without flowers a key has been included (PART G) to deal with water plants solely on a vegetative basis; though it may only be possible to identify a non-flowering aquatic to a genus and not to a species.

Remember that variation in the depth or movement of the water may cause equal variation in the growth form of the plants.

Collecting for identification In order to identify a plant it is usually essential to examine several different parts. If identification can only be done at home it will be necessary to take a radical leaf, one or two stem leaves, a flower and a fruit (or perhaps a spray of flowers or fruit). This should be sufficient, and except for some tiny annuals it should never be necessary to uproot a plant in order to identify it.

The 'Code of Conduct for the Conservation of Wild Plants' adopted by a number of conservation groups offers sound advice. Perhaps the most important is, 'It is illegal for anyone, without permission of the owner or occupier, to uproot any wild plant'.

English names Many plants, especially those without showy flowers, have never acquired a common English name. On the other hand there are many widespread or well-loved plants which have all too many. In this book the name given to every plant is that recommended by The Botanical Society of the British Isles in *English Names of Wild Flowers* by Dony, Perring and Rob, 1974.

4 **Equipment** A hand lens, between ×4 and ×10 magnification, will be a great help when using a few parts of the key. Also useful are a pair of forceps and a needle mounted in a handle.

A rule will be found on the back cover.

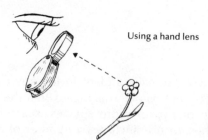

Using a hand lens

Hold the lens close to the eye and in a good light move the specimen towards the lens.

Hybrids

In a number of genera cross-pollination between different species occurs commonly or sporadically. If the hybrid offspring re-cross or back-cross with the parent generation then a widely varying assortment of plants may be found. It has not been possible to include very many of these hybrids in the key.

The following genera are specially liable to hybridization and variation:

Dactylorhiza	Mentha	Salix
Epilobium	Potamogeton	Viola
Festuca	Rosa	

Latin names

Every plant has a Latin name consisting of two words. The first (the genus) may be likened to a surname, and the second (the species) to a descriptive or nickname. The names used in these keys are those used in *Excursion Flora of the British Isles* 1981 (which in turn follows the nomenclature of *Flora Europaea*.

Plants are grouped in families which are here arranged in the same order as in that work.

Measurements

If the key does not state in which direction a flower or leaf is to be measured, then it is intended that the largest measurement should be taken.

Non-flowering plants

These fall into two categories: (1) Plants which do not flower but reproduce by spores (e.g. Ferns). For these see PART A page 17. (2) Plants whose flowers are temporarily absent. Among these many trees and aquatics are fully keyed out using vegetative characters only.

Rarities Great rarities and extremely local plants do not find a place on the keys. These omissions represent only a tiny fraction of our flora, and are seldom found by chance. Plants which only grow in Ireland are also omitted.

A number of plants are marked with an asterisk (*). This means that the species is comparatively rare, or very limited in distribution, or is probably a garden escape.

Simplified keys A few common groups of plants attract more than their fair share of neglect because they are thought to be difficult to identify. Special Simplified Keys on pages 251 to 262 have been provided for these groups to encourge the beginner to tackle them. They include common Ferns, Cow Parsleys, Dandelion-like flowers, Deadnettles and Docks.

For further identification

For more advanced work and comprehensiveness the following manuals are standard reference works:

Flora of the British Isles Clapham, Tutin & Warburg, Cambridge University Press.

Excursion Flora Clapham, Tutin & Warburg, Cambridge University Press.

For difficult groups with full keys and illustrations the following Handbooks are outstanding:

Sedges B.S.B.I. Handbook No. 1
Umbellifers B.S.B.I. Handbook No. 2
Docks & Knotweeds B.S.B.I. Handbook No. 3
Willows & Poplars B.S.B.I. Handbook No. 4
Grasses C. E. Hubbard, Penguin.

The parts of a plant

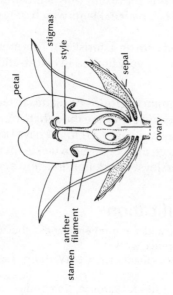

stigmas
style
petal
sepal
stamen
anther
filament
ovary

A flower in detail

The petals collectively form the corolla. The sepals collectively form the calyx. Together they form the perianth. Where the parts of the perianth all resemble each other (i.e. they are not distinguishable as petals and sepals) they are called perianth segments, abbreviated in the key to per. seg.

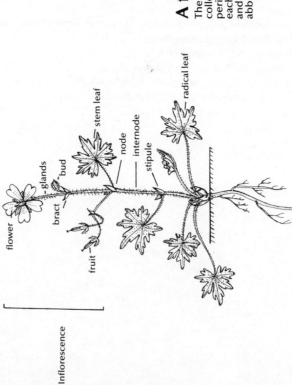

flower
glands
bract
bud
stem leaf
node
internode
stipule
radical leaf
fruit

Inflorescence

Terms used for describing leaves and fruits

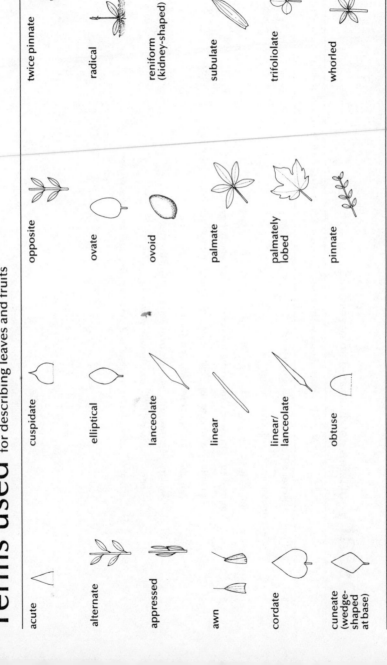

acute

alternate

appressed

awn

cordate

cuneate (wedge-shaped at base)

cuspidate

elliptical

lanceolate

linear

linear/lanceolate

obtuse

opposite

ovate

ovoid

palmate

palmately lobed

pinnate

twice pinnate

radical

reniform (kidney-shaped)

subulate

trifoliolate

whorled

Glossary of terms not illustrated on pages 6 or 7

Special terms used only with certain groups of plants are explained under the appropriate family key

regular

appressed
capsules

auricle

axillary
spike

capsules

Achene   a small one-seeded fruit

Actinomorphic   radially symmetrical

Aggregate   a group of closely related species. See note on page 0.

Appressed   pressed close to the stem or surface on which they grow

Aromatic   with a strong scent. The leaves should be crused to determine this.

Auricle   a lobe at the lower end of a leaf

Axillary   growing in the angle between a leaf and the stem

Berry   a fleshy fruit, usually with several seeds

Bract   often like a very small leaf, on a stem or under a head of flowers

Bulbil   a tiny bulb among the flowers or leaves

Calcareous   chalky or limy

Capsule   a fruit, not fleshy, often round or bottleshaped, but may be long and
                narrow

catkins

decurrent

free

gland

Catkin	a spike of tiny flowers. It may be erect or hanging
Compound	divided into a number of leaflets. A compound leaf is distinguished from a stem with small leaves because here there are no axillary buds.
Deciduous	bare of leaves in the winter
Decurrent	running down the edge of a stalk
Dioecious	bearing male and female flowers (i.e. stamens & stigmas) on separate plants
Entire	without any teeth or intentations along the edge
Floret	one of the tiny flowers that make up the head of a daisy, thistle, grass, etc.
Free	not joined to each other, even at the base
Gland	a tiny shiny globule, sometimes on the end of a hair; found on stems, leaves, sepals etc.
Glaucous	bluish green

Inflorescence   the whole arrangement of flowers on a plant, or sometimes (on trees) a spray of flowers

Internode   the part of a stem between two nodes

internode

ligule

node

pappus

Ligule   part of a grass or sedge leaf

Mealy   covered with a fine, soft, whitish powder

Node   the point on a stem at which a leaf arises, often marked by a ridge round the stem

Pappus   the tuft of hairs on thistle seeds etc., like a parachute

Pinnate   used of a leaf whose leaflets are arranged in opposite pairs

once pinnate   twice pinnate   3 times pinnate

Radical	rising from the base of a stem
Rhizome	an underground stem
Simple	used of a leaf etc. which has no leaflets or large indentations

Solitary	only one on each stem
Spike	an arrangement of flowers set close to the stem, either upright or drooping
Spore	minute reproductive body, one-celled, non-sexual, produced by ferns, horsetails, etc. (These groups do not produce seeds)
Stipule	a small leaflet at the base of a leaf stalk
Striate	used of a stem marked with lines but not grooves or ridges
Succulent	thick and fleshy
Terminal	at the end of a stem
Umbel	an arrangement of stalks radiating like the spokes of an umbrella (See also page 125)

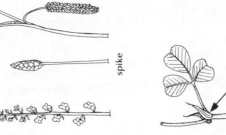

radical

spike

stipule

United   used of sepals or petals joined to each other even if only at their base

Wing   a thin border to a seed or along the angle or a stem

Zygomorphic   not radially symmetrical, e.g. flowers of Pea, Orchid, Deadnettle

united   winged seed   winged stem   zygomorphic

Abbreviations

agg.	aggregate	inflor.	inflorescence
approx.	approximately	lf.	leaf
consid.	considerably	lfless, lflet, lfy	leafless, leaflet, leafy
conspic.	conspicuous	lvs.	leaves
fl., fls.	flower, flowers	mt.	mountain
fr.	fruit	per. segs	perianth segments
hd., hds.	head, heads		

sl.	slightly	w'out	without
sp'let	spikelet	N.	North
succ.	succulent	S.	South
usu.	usually	E.	East
v.	very	W.	West
w.	with	*	uncommon or garden escape

First Key

For meaning of terms see pages 6 to 12.
For abbreviations see page 13.

Part/page

Non-flowering plants

which reproduce by spores
(minute non-sexual bodies)

A/17

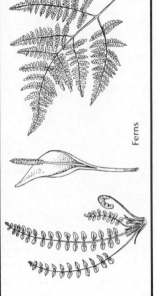

Horsetails

Clubmosses

Ferns

Flowering plants and conifers herbs, shrubs, trees

Plants w. woody stems
(trees, shrubs and undershrubs)

tree w. a trunk [a], in leaf   B/18

shrub, undershrub, or woody climber [b, c], in leaf   C/21

tree or shrub [a, b, c] with fls. appearing before lvs.   D/26

a
tree

b
shrub

c
undershrub

Aquatic plants (growing in permanent water) but not including bog plants

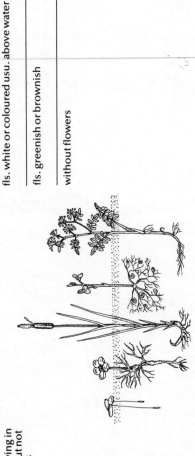

fls. white or coloured usu. above water	E/27
fls. greenish or brownish	F/29
without flowers	G/31

Individual fls. usu. under 5 mm, greenish to brownish, often inconspicuous.
All grasses, sedges and rushes fall into this section

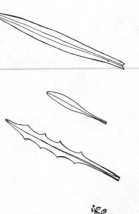

lvs. narrow or grass-like or stem-like, or absent; length more than 4 times width	H/35
lvs. otherwise	J/37

▶

K/41

Many white or coloured fls. or petals packed into one or more tight heads resembling a single flower, e.g. Daisy, Clover, Thistle

L/43

Flowers zygomorphic

M/45

Fls. w. sepals or petals but not both, or sepals may resemble petals

N/47

petals joined to each other at least at base [a]

O/50

petals quite free from each other [b]

Fls. w. distinct sepals & petals

Keys to families

Part A Non-flowering plants, reproducing by spores e.g. Ferns, Horsetails *Note:* The 'leaves' of Ferns are properly called fronds

No true lvs.; stems jointed, often in whorls		spore cases in a terminal cone	EQUISETACEAE 57
Aquatic plants (these are also included in PART G)	whole plant floating on the surface	stem branches, w. minute overlapping lvs.	AZOLLACEAE 65
		lvs. single, or a few joined together	LEMNACEAE (flowering plants whose fls. are not likely to be found) 217
	plant rooted to the bottom	tufted plant; lvs. stiff. cyl. 8–20 cm	ISOETACEAE 57
		creeping; lvs. 3–15 cm; spore cases pill-like	MARSILEACEAE 65
Lvs. not much over 6 mm, closely set; plant often like a robust moss	lvs. packed all round the stem		LYCOPODIACEAE & SELAGINELLACEAE 56
	frond flattened	lvs. 1–3 mm long, ovate	SELAGINELLACEAE 56
		frond up to 10 cm w. translucent segments	HYMENOPHYLLACEAE 59
Spore cases in spikes or clusters above the lvs.	plant under 20 cm high w. only 1 or 2 lvs.		OPHIOGLOSSACEAE 65
	plant 50–300 cm high		OSMUNDACEAE 59

17

▶

Spore cases in small capsules among lvs.	lvs. translucent		HYMENOPHYLLACEAE	59
Other fern-like plants	fronds barren or w. spores on back		various Fern families	60

Part B Trees in leaf

Lvs. under 5 mm, almost scale-like	trees with woody cones under 15 mm		CONIFERAE	66
	seaside shrub with pink fls.		TAMARICACEAE	86
Lvs. needle-like, or small and linear			CONIFERAE	66
Lvs. compound w. separate lflets	lvs. opposite	lvs. palmate	HIPPOCASTANACEAE	101
		buds black	OLEACEAE	152
		lvs. pinnate	CAPRIFOLIACEAE	176
	lvs. alternate, pinnate		ROSACEAE	110
Lvs. opposite and 3–7 lobed	lvs. w. narrow stipules	fls. white	CAPRIFOLIACEAE	176
	no stipules	fls. greenish	ACERACEAE	101
Lvs. opposite simple	lvs. finely toothed	branches usu. spiny; lvs. w. 3 pairs of veins	RHAMNACEAE	102
		not spiny; lvs. w. 7–10 pairs of veins	CELASTRACEAE	102

lvs. cordate or lanceolate	lvs. thick, shiny, up to 25 mm; fls. green	BUXACEAE	102
	lvs. over 25 mm; fls. white or lilac	OLEACEAE	152
Many lvs. with strong spines	lvs. flat, dark green above, whitish below	FAGACEAE	139
	lvs. wavy	AQUIFOLIACEAE	102
Lvs. lobed	a small thorny tree	ROSACEAE	110
	lvs. palmately veined	SALICACEAE	140
	lvs. green below; fr. spherical	PLATANACEAE	137
	buds brown	FAGACEAE	139
	buds green	ROSACEAE	110
Lf. buds w. only one scale		SALICACEAE	140
Branches spiny		ROSACEAE	110
Lf. margins entire but sometimes wavy	fls. white, showy; stamens many	ROSACEAE	110
	lf. buds w. many scales	FAGACEAE	139
	lf. buds w'out scales	small tree	RHAMNACEAE	102

19

▶

Young lvs. downy beneath, at least on the veins	If. base asymmetrical; lvs. rough		ULMACEAE	137
	lvs. thick, dark; evergren	tree, bearing nuts in a cup (acorns)	FAGACEAE	139
	bearing woody cone-like catkins		BETULACEAE	138
	fls. in catkins (which may appear before lvs.)	tree	SALICACEAE	140
		large shrub; lvs. almost as wide as long	CORYLACEAE	138
	inflor. stalk ½ joined to large bract; fls. cream	large tree w. heart-shaped lvs.	TILIACEAE	96
	fls. w. white petals	smaller tree	ROSACEAE	110
Lvs. whitish beneath, or stalks flattened	catkins appear before lvs.		SALICACEAE	140
Lvs. cordate, ♡ toothed	lvs. w. large and small teeth	shrub or small tree	CORYLACEAE	138
	If. teeth regular	large tree	TILIACEAE	96
If. base asymmetrical; lvs rough			ULMACEAE	137
Lvs. hardly longer than wide	tree w. some woody cone-like catkins	If. tip obtuse ⌒	BETULACEAE	138
	shrub, bearing nuts		CORYLACEAE	138
	tree		SALICACEAE	140

Usu. 2 knob-like glands near top of lf. stalk			ROSACEAE	110
Lvs. toothed	lvs. 10–25 cm long		FAGACEAE	139
	If. base obtuse or cordate	lvs. 3–10 cm	CORYLACEAE	138
	If. base usu. obtuse angled	lvs. usu. under 5 cm	BETULACEAE	138

Part C Shrubs and undershrubs in leaf Note: If the specimen being examined is thought likely to grow into a tree, look also in Part A.

Lvs. under 4mm, scale-like, overlapping	up to 60 cm high		ERICACEAE	146
	1–3 m high		TAMARICACEAE	86
Lvs. v. succulent, 3-angled	fls. many-petalled, showy	S.W. coasts	AIZOACEAE	93
Climbing or scrambling shrub, not prickly or spiny	lvs. alternate	evergreen; main stem covered w. rootlets	ARALIACEAE	125
		fls. purple; berries red	SOLANACEAE	158
	lvs. compound		RANUNCULACEAE	68
	lvs. simple, opposite		CAPRIFOLIACEAE	176
Parasite, usu. high on a tree	lvs. in term. pairs, evergreen		LORANTHACEAE	124
	lvs. opposite		CAPRIFOLIACEAE	176
Lvs. pinnate w. 3 to many lflets.			▶	21

lvs. alternate	fls. pea-like	LEGUMINOSAE	103
	fls. white/pink; over 12 stamens	ROSACEAE	110
	lvs. spiny, w'out stipules	BERBERIDACEAE	71
Lvs. up to 2 cm, in whorls of 3 or more, or closely crowded	lvs. spine-tipped, in whorls of 3	fls. yellowish; fr. berry-like, blue-black	CUPRESSACEAE	66
	petals united; corolla bell-shaped	heath plants	ERICACEAE	146
	petals free, usu. 5; stamens usu. 6	near saltmarshes; lvs. under 5 mm	FRANKENIACEAE	87
Lvs. palmate or palmately lobed. 	lvs. w. 7–10 lflets.		LEGUMINOSAE	103
	lvs. 3–7 lobed	fls. white; inflor. 5–10 cm across	CAPRIFOLIACEAE	176
		lvs. up to 10 cm; stamens 5	GROSSULARIACEAE	118
		lvs. up to 20 cm; stamens over 12; S. & W. coasts	MALVACEAE	97
Plant spiny	lvs. reduced to spines	fls. yellow	LEGUMINOSAE	103
	lvs. lobed	shrub up to 1 m high; stamens 5	GROSSULARIACEAE	118
		2–8 m high; more than 12 stamens per fl.	ROSACEAE	110
	spines 3-pronged; lvs. in clusters		BERBERIDACEAE	71

lvs. evergreen, w. a spiny tip	under 1 m high	LILIACEAE	202
lvs. silvery, linear-lanceolate, lanceolate, up to 8 cm	usu. maritime, fr. orange	ELAEAGNACEAE	120
fls. white, w. 5 petals	fls. appear before lvs.	ROSACEAE	110
fls. of the Pea type, yellow or pinkish [a]		LEGUMINOSAE	103
fls. purplish		SOLANACEAE	158
fls. greenish white w. 4 petals	lvs. ovate, toothed, 3–6 cm	RHAMNACEAE	102
lvs. opposite or nearly so	lvs. palmately lobed	inflor. flat-topped, white; fr. a berry	CAPRIFOLIACEAE	176
		inflor. greenish; fr. dry, winged	ACERACEAE	101
	lvs. toothed, sometimes v. finely	lvs. lanceolate, 10–25 cm fls. purple	BUDDLEJACEAE	151
		lvs. usu. downy below; inflor. flat, 6–10 cm	CAPRIFOLIACEAE	176
		fls. red & purple, drooping	ONAGRACEAE	120
		lvs. 3–13 cm; twigs green; fr. pink/orange	CELASTRACEAE	102
		lvs. 3–6 cm; twigs brown; fr. green to black	RHAMNACEAE	102

▶

lvs. 5–10 cm long	fls. yellow; stamens over 12	HYPERICACEAE	105
	fls. white; stamens 4	CORNACEAE	124
	fls. lilac or white; stamens 2	OLEACEAE	152
fls. yellow, 1–2 cm across		CISTACEAE	86
lvs. up to 1 cm; fls. pink to purple; low, creeping plant	corolla 4-lobed; common	LABIATAE	166
	corolla 5-lobed; stamens 5; N. Scotland	ERICACEAE	146
	petals 5, crinkly; stamens 6; near S. & E.	FRANKENIACEAE	87
lvs. thick, shiny, up to 25 mm	fls. greenish, in Spring	BUXACEAE	102
lvs. 2–6 cm	fls. white	OLEACEAE	152
	fls. pink; fr. white	CAPRIFOLIACEAE	176
Lvs. evergreen w. spiny tip		LILIACEAE	202
Some lvs. w. 3 lflets		LEGUMINOSAE	103
Upper lvs. deeply lobed, silvery		COMPOSITAE	179
Lvs. w. teeth (sometimes only a few small ones near tip)	buds w. only one covering scale	SALICACEAE	140

thorny bush; fs. about 4 mm, greenish; petals 4	fr. berry-like, green to black	RHAMNACEAE	102
If. teeth sharp, sometimes in 2 sizes	lvs. lanceolate; inflor. dense, pink	ROSACEAE	110
	fls. in drooping catkins	CORYLACEAE	138
fls. globular to bell-shaped; fr. a berry	plant up to 1 m high	ERICACEAE	146
lvs. broad, up to 15 mm		BETULACEAE	138
If. buds w. only 1 scale	catkins green/yellow opening w. or after lvs.	SALICACEAE	140
If. buds w. sev. scales; twigs reddish	catkins reddish (usu. opening before lvs.)	MYRICACEAE	137
Lvs. dark, glossy, evergreen	fls. w. 4 sepals, no petals, 8 stamens	THYMELAEACEAE	120
	corolla 5-lobed; stamens 10	ERICACEAE	146
Fls. yellow zygomorphic, pea type	lvs. small, narrowly lanceolate	LEGUMINOSAE	103
Buds hairy, w'out scales; fr. berry-like	shrub or small tree	RHAMNACEAE	102
Buds w. only one covering scale		SALICACEAE	140
Lvs. w. stipules	fls. white/pink; stamens over 12	ROSACEAE	110
Fls. globular, pinkish	moors & bogs, mostly in N.	ERICACEAE	146

▶

	perianth 6-lobed, stamens 0 or 3	moor & mt. plants; fr. black	EMPETRACEAE	149
Lvs. aromatic; fls. in short catkins			MYRICACEAE	137
Fls. pink/purple w. protruding stamens	petals 4		ERICACEAE	146
	petals 5		SOLANACEAE	158
	moorland plants		ERICACEAE	146
	maritime plants		CHENOPODIACEAE	93

Part D Trees and shrubs whose flowers appear before their leaves

Fls. white w. more than 12 stamens			ROSACEAE	110
Usu. thorny shrubs; fls. greenish, about 5 mm in small clusters	stamens 0 or 4	coastal	ELAEAGNACEAE	120
Fls. in dangling catkins	some catkins woody, ovoid, erect		BETULACEAE	138
	If. buds fat, blunt	large shrub	CORYLACEAE	138
	If. buds pointed, often narrow, w. only 1 scale	tree	SALICACEAE	140
Buds black, in pairs			OLEACEAE	152
Fls. in erect catkins; shrub or small tree	aromatic shrub; catkins reddish; stamens 0 or 4		MYRICACEAE	137

			Family	Page
catkins green or yellow	lf. buds w. only 1 outer scale		SALICACEAE	140
Large tree			ULMACEAE	137

Part E Aquatic plants with aerial white or coloured flowers

				Family	Page
Floating lvs. round or oval, or elliptical, hardly lobed	fls. w. 3 petals	lvs. elliptical		ALISMATACEAE	195
		floating plant w. round lvs.		HYDROCHARITACEAE	196
	sepals, petals & stamens 5			MENYANTHACEAE	154
	sepals 4-6; petals & stamens over 12			NYMPHAEACEAE	72
Fls. in umbels, which may be minute (only 5 mm across)	fls. w. 3 petals	lvs. 3-angled; plant over 40 cm		BUTOMACEAE	196
		lvs. w. flat blade; plant up to 20 cm		ALISMATACEAE	195
	fls. w. 5 petals			UMBELLIFERAE	125
Lvs. in whorls of 3 to 5	lvs. v. finely divided	fls. lilac		PRIMULACEAE	150
		fls. yellow, showy		LENTIBULARIACEAE	165
		fls. small, in a spike		HALORAGACEAE	122
	lvs. small, in 3's			HYDROCHARITACEAE	196
▶					27

	fls. showy, purple	LYTHRACEAE	119
Lvs. w. 3 lflets.		MENYANTHACEAE	154
Fls. zygomorphic	Fls. yellow	LENTIBULARIACEAE	165
	fls. blue/lilac	LOBELIACEAE	174
Fls. under 5 mm, pink/green in a spike		POLYGONACEAE	133
Fls. w. 3 petals	plant floating, not rooted	HYDROCHARITACEAE	196
	tiny creeping plant	ELATINACEAE	87
	plant upright or floating & rooted	ALISMATACEAE	195
Fls. w. 4 sepals and 6 stamens		CRUCIFERAE	75
Fls. w. 5 petals or 5 large yellow sepals	stamens 4 — tiny creeping plant	SCROPHULARIACEAE	159
	stamens 5 — lvs. linear / in a rosette	CAMPANULACEAE	173
	fls. white	UMBELLIFERAE	125
	fls. lilac	PRIMULACEAE	150
	fls. yellow; petals w. a fringe	MENYANTHACEAE	154

stamens more than 12	fls. 4–6 cm across, yellow; lvs. 10–30 cm, oval/round	NYMPHAEACEAE	72
	fls. white	RANUNCULACEAE	68
Fls. red/purple w. 6 petals	stamens 12	LYTHRACEAE	119

Part F Aquatic plants with small greenish or brownish flowers either aerial or submerged

Fls. in flat cluster on grass-like stem	marine plants below high tide level	ZOSTERACEAE	198
Lvs. in whorls	spores produced in a brownish terminal cone	EQUISETACEAE	57
	fls. floating on slender stalks	HYDROCHARITACEAE	196
	whole plant submerged	CERATOPHYLLACEAE	72
	lvs. linear, 6–12 in a whorl [a]	HIPPURIDACEAE	123
	lvs. pinnate, 4–5 in a whorl [b]	HALORAGACEAE	122
Fls. tightly packed in a large cylindrical spike	spike projecting from side of stem	ARACEAE	216
	spike upright, terminal	TYPHACEAE	218
Lvs. v. fine, all submerged or floating	fr. hardly stalked, in lf. axils	ZANNICHELLIACEAE	202
	fr. in simple umbel on a long stalk	in coastal ponds and ditches	RUPPIACEAE	201
	▶		29

a

b

Fls. in globular or starry heads	fls. w. 4 sepals; lvs. w. stipules	POTAMOGETONACEAE	198
	fls. in sedge-like spike	CYPERACEAE	219
lvs. linear		SPARGANIACEAE	218
Large dock-like plant		POLYGONACEAE	133
All lvs. submerged and/or floating — tiny surface plants; roots hanging or absent	floating fronds under 10 mm	LEMNACEAE	217
lvs. linear / in a basal tuft	4 sepals; 6 stamens	CRUCIFERAE	75
fls. in spikes on stalks		POTAMOGETONACEAE	198
fls. not stalked, axillary		CALLITRICHACEAE	123
Lf. veins branching		POLYGONACEAE	133
Fls. w. 6 sepals		JUNCACEAE	206
Stems hollow lvs. linear — lvs. w. saw-edge keel		CYPERACEAE	219
lvs. mostly flat; stem w. nodes	grasses	GRAMINEAE	230
Other plants w. linear leaves	sedges	CYPERACEAE	219

Part G Aquatic plants without flowers

Note: Towards the end of this part of the key are a number of plants (of different families) which cannot easily be keyed out on vegetative characters alone. It is worth searching carefully for the flowers or fruit (often very inconspicuous) as the plants can then more easily be placed in their correct family by using Parts E or F of the key.

Plant grass-like, below high-tide level	lvs. linear	ZOSTERACEAE	198
Plants floating at surface, not rooted to the bottom	lvs. lanceolate, stiff, toothed, 15–50 cm	HYDROCHARITACEAE	196
	lvs. rounded, about 3 cm across	HYDROCHARITACEAE	196
	lvs. about 1 mm, overlapping along branched stem / stem rises above surface	AZOLLACEAE	65
	plant up to 12 mm across, usu. oval or lobed	LEMNACEAE	217
Plants w. broad lvs. (ovate or round) which float on surface, but no aerial leaves	lvs. roundish, cordate, not lobed / lvs. 3–10 cm across	MENYANTHACEAE	154
	lvs. 10–30 cm across	NYMPHAEACEAE	72
	If. outline round, 1–3 cm across, usu. lobed / finely divided submerged lvs. sometimes present	RANUNCULACEAE	68
	lvs. up to 3 cm, upper ones in a rosette	CALLITRICHACEAE	123
	lvs. w. only one long vein / floating lvs. 5–15 cm long	POLYGONACEAE	133
	lvs. w. stipules	POTAMOGETONACEAE	198

▶

31

Stems w. lobed or pinnate lvs. rising above the surface	lf. base tapering or almost cordate		ALISMATACEAE 195
	upper lvs. w. 2 pointed lobes at base		ALISMATACEAE 195
	lvs. w. 3 lflets.		MENYANTHACEAE 154
	lvs. in whorls, finely pinnate [a]		HALORAGACEAE 122
	lvs. pinnate	lflets toothed, pointed	UMBELLIFERAE 125
		lflets only bluntly indented	CRUCIFERAE 75
Lvs. ovate w. entire margins, aerial			ALISMATACEAE 195
Aerial lvs. in whorls, narrow	stem jointed, hollow		EQUISETACEAE 57
	lvs. flat, linear		HIPPURIDACEAE 123
Stems rising above surface w. linear aerial lvs.	stem hollow w. nodes		GRAMINEAE 230
	lvs. 3-angled	lvs. all basal, X-section triangular	BUTOMACEAE 196
		lvs. with conspic. veins; X-section triangular	SPARGANIACEAE 218
		lvs. keeled to flat; fl. stem often triangular	CYPERACEAE 219
	lvs. flat, opposite	plant 1–2.5 m high	TYPHACEAE 218

a

Stem rising above surface, lfless	inflor. a brown spike		CYPERACEAE	219
Lvs. submerged, pinnate or finely forked	small bladders usu. present among some lvs.		LENTIBULARIACEAE	165
	lvs. in whorls, clearly pinnate	lvs. up to 4.5 cm long [a]	HALORAGACEAE	122
		lvs. 5–10 cm long	PRIMULACEAE	150
	lvs. alternate pinnate w. narrow segments		UMBELLIFERAE	125
	lvs. in whorls, forked, toothed, stiff	tip of plant often tassel-like	CERATOPHYLLACEAE	72
	lvs. variously and finely divided		RANUNCULACEAE	68
Lvs. lanceolate, stiff, clearly toothed			HYDROCHARITACEAE	196
Lvs. in whorls, all submerged	whorls of 6–12 linear lvs.		HIPPURIDACEAE	123
	whorls of 3 oblong lvs.		HYDROCHARITACEAE	196
Small creeping mud plant	lvs. in pairs	term. lvs. in a rosette; no stipules	CALLITRICHACEAE	123
		lvs. w. tiny (about 2 mm) stipules; uncommon	ELATINACEAE	87
Plants w. a small rooted tuft of linear lvs. 2–20 cm long See NOTE on page 00	sometimes forms a turf underwater	lvs. often ½ cylindrical	PLANTAGINACEAE	172
	young lvs. w. coiled tip	creeping plant	MARSILEACEAE	65

▶ 33

Group	Feature	Condition	Family	Page
	cross veins clearly visible near lf. base	in N. & W.	ISOETACEAE	57
	lvs. w. 2 interior tubes	in N. & W.	LOBELIACEAE	174
	lvs. cylindrical	in acid water of N. & W.	CRUCIFERAE	75
Lower lvs. semi-cylindrical or 3-angled			SPARGANIACEAE	218
Lvs. linear, mostly floating See NOTE on page 31	inflor. globular, mostly sessile		SPARGANIACEAE	218
	inflor. a small long-stalked spike		CYPERACEAE	219
Lvs. mostly submerged See NOTE on page 31	lvs. well over 6 mm wide		POTAMOGETONACEAE	198
	lvs. mostly opposite	lf. tip notched	CALLITRICHACEAE	123
		lf. tip finely pointed	ZANNICHELLIACEAE	202
	lvs. usu. w. a stipule		POTAMOGETONACEAE	198
	in brackish water		RUPPIACEAE	201
	in fresh water		CYPERACEAE	219
Other plants w. fine or linear lvs.			POTAMOGETONACEAE	198

Part H Land plants with small greenish or brownish flowers and narrow leaves typically more than 4 times as long as wide

Fresh stems w. sticky, milky juice			EUPHORBIACEAE 131
Lower lvs. once or twice pinnate	stem lfless; fls. in a spike		PLANTAGINACEAE 172
	lvs. greyish/woolly on at least one side		COMPOSITAE 179
	stamens 2–6		CRUCIFERAE 75
Lvs. in whorls round the stem			EQUISETACEAE 57
Plant w. onion smell			LILIACEAE 202
Seaside plants w. succulent leaves	lvs. all radical, linear fls. in a spike	fls. w. stalks and 6 perianth segments	JUNCAGINACEAE 197
		fls. not stalked; sepals 4	PLANTAGINACEAE 172
	lvs. various		CHENOPODIACEAE 93
Inflor. in spikes; stamens 12 or more	petals variously lobed		RESEDACEAE 83
Lvs. greyish, downy, but not grasslike			COMPOSITAE 179
▶			35

			Family	Page
Petals and/or sepals in 4's	lvs. opposite		CARYOPHYLLACEAE	87
	lvs. all radical		PLANTAGINACEAE	172
Petals and/or sepals in 5's	lvs. opp.	stigmas 3; Scottish mts.	CARYOPHYLLACEAE	87
		fr. enclosed by 2 bracts	CHENOPODIACEAE	93
		lvs. linear, 5–15 mm long	CARYOPHYLLACEAE	87
	sepals & petals present		RANUNCULACEAE	68
	creeping plant; 3 bracts below ea. fl.	in calcareous grassland	SANTALACEAE	124
	plants of shingle or saltmarsh		CHENOPODIACEAE	93
	fls. 1–3 in cluster w. narrow bracts beneath		UMBELLIFERAE	125
6 sepals per fl.	fl. spike dense, oblique; lvs. stout, crinkled		ARACEAE	216
	fls. in a long, narrow erect spike		JUNCAGINACEAE	197
	fls. in clusters, or inflor. loose	stamens 3 or 6	JUNCACEAE	206
	lvs. succulent, or stiff and up to 4 cm	maritime plants	CHENOPODIACEAE	93
Stem hollow, w. distinct nodes, lvs. linear	lvs. w. a toothed keel		CYPERACEAE	219

	grasses; stamens and stigmas 1–3	stem cross-section round or elliptical	GRAMINEAE 230
Stem. usu. solid; nodes not distinct; lvs. linear	stamens & stigmas 2–3; sedges and similar	stem round or elliptical or triangular	CYPERACEAE 219

Part J Plants with small greenish or minute flowers, and leaves that are *not* long and narrow (typically length less than 4 times width)

Stinging plant			URTICACEAE 136
Fl. hd. like a daisy w' out rays	lvs. finely dissected		COMPOSITAE 179
Lvs. 3-lobed, clover-like	fls. actinomorphic, 5 sepals and 5 or 0 tiny petals	on Northern hills only	ROSACEAE 110
	fls. zygomorphic, of Pea type		LEGUMINOSAE 103
Climbing, twining, or scrambling plant	lvs. alternate, heart-shaped or 3-lobed	calyx and fr. 3-lobed	POLYGONACEAE 133
		lvs. shiny; fr. a berry	DIOSCOREACEAE 210
	spiral tendrils opposite lvs.		CUCURBITACEAE 130
	lvs. opposite		CANNABACEAE 136
	lvs. in whorls of 4–6		RUBIACEAE 174
Club-like spike inside a leafy sheath	fls. at base of club		ARACEAE 216

37

▶

			Family	Page
Obvious milky juice in fresh stems			EUPHORBIACEAE	131
Few or no lvs. on fl. stem	lvs. round, joined to stalk by centre	creeping plant of damp places	UMBELLIFERAE	125
	succulent plant of tidal mud		CHENOPODIACEAE	93
	fls. in 5's at top of stem, greenish [a]	terminal fl. w. 4 petals, others w. 5	ADOXACEAE	177
	lvs. once or twice pinnate	lvs. twice pinnate; stamens v. prominent	RANUNCULACEAE	68
		lf. lobes broad, toothed	ROSACEAE	110
		lf. lobes linear	PLANTAGINACEAE	172
	lvs. palmate		ROSACEAE	110
	lvs. kidney-shaped		POLYGONACEAE	133
	inflor. a long-stalked spike	bog plant, usu. aquatic	POTAMOGETONACEAE	198
		lvs. all radical	PLANTAGINACEAE	172
		lvs. opposite	EUPHORBIACEAE	131
Fls. in 5's at top of stem	lvs. & lflets usu. 3-lobed		ADOXACEAE	177
V. low creeping plant on mud or damp walls or marshes	lvs. lobed to pinnate		CRUCIFERAE	75

	lvs. opposite, linear to ovate		✿	CALLITRICHACEAE	123
	lvs. alternate, almost round	lvs. up to 6 mm across; calyx 4-lobed	/	URTICACEAE	136
		lvs. up to 2 cm; fls. in ones	○	SCROPHULARIACEAE	159
		lvs. mostly 2–5 cm, joined to stalk at centre	◗	UMBELLIFERAE	125
Fls. in umbels	4 sepals or petals per fl.	lvs. broad		ROSACEAE	110
	5 sepals or petals per fl.			UMBELLIFERAE	125
Seaside plant, either shrubby or prickly				CHENOPODIACEAE	93
l.f. bordered w. thin black line beneath	under 5 cm tall			PRIMULACEAE	150
Lvs. opposite	corolla w. 2 large and 3 small lobes	fls. purplish brown		SCROPHULARIACEAE	159
	3 sepals per fl.			EUPHORBIACEAE	131
✿	sepals 3-toothed; stamens 4			LINACEAE	97
	fr. enclosed by 2 bracts			CHENOPODIACEAE	93
	lvs. stalked, ovate; ○ seeds black, visible			CHENOPODIACEAE	93
	calyx 6 or 12 toothed			LYTHRACEAE	119
▶					39

	lvs. simple, margins entire	usu. in dry, open places	CARYOPHYLLACEAE	87
	lvs. rounded, w. shallow lobes	usu. in damp, shady places	SAXIFRAGACEAE	117
Inflor. a robust spike; stamens more than 10, showy	petals lobed, yellowish; lvs. pinnate, wavy		RESEDACEAE	83
Fr. 3-sided	3 small and/or 3 large sepals	Docks, etc.	POLYGONACEAE	133
8–10 stamens per fl.	low creeping plant of damp shady places	lvs. roundish w. shallow lobes	SAXIFRAGACEAE	117
Plant forming a low dense mat	lvs. almost round, up to 6 mm	usu. on walls	URTICACEAE	136
Lvs. greyish, woolly or mealy on at least one surface	fls. in yellowish brown clusters		COMPOSITAE	179
	fls. greenish, in spikes	fl. spikes dense, mixed w. many bristles	AMARANTHACEAE	93
		fls. in clusters or small spikes	CHENOPODIACEAE	93
Strongly smelling plants	lower lvs. lobed to pinnate		CRUCIFERAE	75
Lvs. once to 3 times pinnate	inflor. a long-stalked knob		ROSACEAE	110
	inflor. looser	stamens more than 12, large	RANUNCULACEAE	68
		2–6 stamens per fl.	CRUCIFERAE	75
Lvs. roundish, lobed and/or toothed			ROSACEAE	110

Lvs. lanceolate, margins entire; 1 style; 4 stamens	stem & perianth reddish	usu. on walls	URTICACEAE	136
Other plants w. small greenish fls.	fls. usu. in small clusters		CHENOPODIACEAE	93

Part K Flowers usually tiny, but packed together into a round or oblong head which may be quite showy, e.g. Thistle, Daisy, Clover, etc.

Climbing plant w'out green lvs.	stems usu. red		CONVOLVULACEAE	157
Plant w. onion smell			LILIACEAE	202
Lvs. v. succ., 3-angled	on S.W. cliffs		AIZOACEAE	93
Lvs. opposite	stems 4-angled; stamens 4; fr. of 4 nutlets	fls. often lilac; lvs. often scented	LABIATAE	166
	fl. hds. Daisy-like w. white or yellow rays		COMPOSITAE	179
	lvs. w. 3 large lflets.	usu. 50–120 cm high; fls. pink/purple	COMPOSITAE	179
	fls. dark purple w. large white bracts	moors in N.	CORNACEAE	124
	petals bluish to lilac	4 stamens per fl.	DIPSACACEAE	178
		2 stamens per fl.	VALERIANACEAE	177
Fl. hd. obviously like a Daisy, Dandelion, or Thistle	each floret w. 5 sepals, 5 petals, 5 stamens	lvs. spiny like Holly, in coastal sand	UMBELLIFERAE	125

▶

no sepals at base of each separate floret		lvs. not spiny; fl. head on a long stalk	CAMPANULACEAE	173
			COMPOSITAE	179
Each tiny fl. of the Pea type 🌿		Clovers, etc.	LEGUMINOSAE	103
Each fl. w. 5 styles		fls. pink or purplish blue	PLUMBAGINACEAE	149
Inflor. maroon; each fl. w 4 sepals & 4 stamens			ROSACEAE	110
Fls. white/pink	lvs. 3-lobed, clover-like 🌿		LEGUMINOSAE	103
	lvs. simple; fl. hds. in a spike		COMPOSITAE	179
		lvs. large, rounded, toothed	POLYGONACEAE	133
		If. margins entire; stamens usu. 8		
		fls. in an umbel; lvs. lobed or compound	UMBELLIFERAE	125
	fls. like whitish 5-petalled daisies in broad compact inflor.	lvs. compound; stem woolly	COMPOSITAE	179
Inflor. a greenish sphere, long-stalked	lvs. pinnate 🌿		ROSACEAE	110
Each fl. w. 5 free stamens	petals blue, long, narrow	style often longer than petals	CAMPANULACEAE	173
Other plants w. small white/coloured fls. packed into a single head	anthers joined in a column	1 style branching into 2 above	COMPOSITAE	179
		2 styles per fl.; inflor. in umbel	UMBELLIFERAE	125

	fls. w. either 4 stigmas or 8 stamens			CRASSULACEAE	115
	fl. heads tightly packed w. tiny fls.			COMPOSITAE	179

Part L Land plants with zygomorphic flowers

Brownish plants w'out green lvs.	each fl. w. 4 stamens and a 2-lobed stigma	lower corolla lip 3-lobed		OROBANCHACEAE	164
	perianth 6-lobed			ORCHIDACEAE	211
Lvs. w. 3 to many distinct lflets	stamens more than 12 (easily seen)	fls. white or yellow; petals small, lobed		RESEDACEAE	83
		fls. 25 mm or more		RANUNCULACEAE	68
	corolla w. a hood and broad 3-lobed lip			SCROPHULARIACEAE	159
	sepals 2, free, up to 6 mm			FUMARIACEAE	74
	sepals 5, usu. joined to calyx tube			LEGUMINOSAE	103
Sepals 2; fl. w. a spur; stamens joined together				BALSAMINACEAE	100
Fls. w. 1 stamen				VALERIANACEAE	177
Stamens 0 or 2 (or apparently 2) or 6	petals 4, of unequal size, blue or pink	lvs. opposite		SCROPHULARIACEAE	159
	fls. solitary on a long stalk; lvs. all radical	bog plants		LENTIBULARIACEAE	165
▶					43

44

	stem w. brownish scales but no normal lvs.		ORCHIDACEAE 211
	lvs. divided into small lobes		FUMARIACEAE 74
	lf. veins parallel		ORCHIDACEAE 211
	lf. veins branching		LABIATAE 166
3 stamens			IRIDACEAE 210
4 stamens	fls. 4 mm across, lilac, in narrow spikes	fls. only slightly zygomorphic	VERBENACEAE 165
	4 nutlets visible; use lens to see inside ripe calyx [a]		LABIATAE 166
	fr. a capsule		SCROPHULARIACEAE 159
5 stamens (which may be joined together)	2 sepals; fl. w. a spur; stamens joined together		BALSAMINACEAE 100
	fl. stalk usu. much longer than fl.	fls. in ones	VIOLACEAE 84
	fls. short-stalked	1 stigma; 4 nutlets visible in ripe calyx [b]	BORAGINACEAE 154
		2 or 3 stigmas	CAMPANULACEAE 173

a

b

8 stamens a	usu. over 50 cm high; petals 4	fls. 2–3 cm, only slightly zygomorphic	ONAGRACEAE	120
	small, dainty plant w. simple lvs.	fls. under 1 cm [a]	POLYGALACEAE	85
Stamens more than 12, easily seen	petals small, lobed, white or yellow		RESEDACEAE	83
	petals few and large		RANUNCULACEAE	68
Stamens 10, lying within keel of lower petal	calyx w. 5 teeth; fl. of Pea type		LEGUMINOSAE	103

Part M Land plants with actinomorphic (radially symmetrical) flowers and a single perianth, i.e. either sepals or petals but not both

Fls. in umbels	petals & stamens 5		UMBELLIFERAE	125
	petals & stamens 6	plant smells of onion	LILACEAE	202
Stamens 3 or 0	corolla 5-lobed	tiny compact Scottish plant; fls. green	CARYOPHYLLACEAE	87
		style long, 3-lobed, fls. pink to lilac	VALERIANACEAE	177
	stem twining w. cordate lvs.	fls. greenish yellow	DIOSCOREACEAE	210
	lvs. linear, over 10 cm	fls. yellow, orange, blue, or purple	IRIDACEAE	210
Stamens 6	sepals 4; petals 0		CRUCIFERAE	75
	▶			45

			Family	Page
	petals 5, pink/white		POLYGONACEAE	133
	perianth segments 6	stem twining w. cordate lvs.	DIOSCOREACEAE	210
		plant smells of onion	LILIACEAE	202
		ovary below corolla e.g. Daffodil, Snowdrop	AMARYLLIDACEAE	209
		ovary inside corolla	LILIACEAE	202
Stamen 1 only; corolla w. long tube			VALERIANACEAE	177
Stamens 8	petals 5, pink/white		POLYGONACEAE	133
	petals & sepals 4, greenish	lvs. 4, in a whorl	TRILLIACEAE	205
Stamens more than 12	petals 4; sepals 2 which fall when fl. opens		PAPAVERACEAE	73
	petals 5 or more		RANUNCULACEAE	68
Lvs. in whorls			RUBIACEAE	174
Styles 2 or 3	lvs. opposite		CARYOPHYLLACEAE	87
	fls. yellowish; lvs. linear w'out stipules		UMBELLIFERAE	125
	fls. green/white/pink; lvs. w. stipules	stipules often tubular	POLYGONACEAE	133

Character	Detail	Note	Family	Page
Fls. in tight brownish hds.			ROSACEAE	110
Fls. w'out stalks, pink, in lf. axils	lvs. ovate, up to 12 mm, usu. opposite	in salty places	PRIMULACEAE	150
Fls. shortly stalked, greenish white	lvs. linear, up to 15 mm, in ones	in calcareous grassland	SANTALACEAE	124

Part N Land plants with actinomorphic (radially symmetrical) flowers, sepals, and united petals

Character	Detail	Note	Family	Page
Stamens more than 15, springing from top of a tube	petals joined to base of tube		MALVACEAE	97
Sepals almost nil; (beware small bracts at fl. base)	lvs. in whorls		RUBIACEAE	174
	lvs. opposite; stamens 0, 1, or 3		VALERIANACEAE	177
	lvs. alternate; stem and perianth reddish		URTICACEAE	136
Sepals 2 or 3	sepals really 5, hidden behind 2 large bracts	climbing plant	CONVOLVULACEAE	157
	sepals 2; lvs. simple, up to 3 cm (not counting stalk)		PORTULACACEAE	92
	lvs. variously lobed	fls. greenish, 5 in a head	ADOXACEAE	177
	lvs. linear		IRIDACEAE	210
Sepals 4	lvs. in whorls		RUBIACEAE	174
	stamens 2; corolla 4-lobed		SCROPHULARIACEAE	159

▶

	stamens 4 or 5	tiny creeping plant, on mud	SCROPHULARIACEAE	159
		lf. margins entire	GENTIANACEAE	153
		lvs. variously lobed	VERBENACEAE	165
Lvs. succ., rounded	stamens 10		CRASSULACEAE	115
Lvs. w. 3 lflets	usu. aquatic		MENYANTHACEAE	154
Lvs. in whorls	fls. white or yellow		PRIMULACEAE	150
	fls. lilac		RUBIACEAE	174
Lvs. radical only	stamens 4; fls. under 3 mm across	growing on wet mud	SCROPHULARIACEAE	159
	stamens 5	5 fine styles	PLUMBAGINACEAE	149
		only 1 style	PRIMULACEAE	150
	stamens 6		AMARYLLIDACEAE	209
Lvs. opposite	lvs. shining, evergreen; margins entire	fls. blue, 25–50 mm across	APOCYNACEAE	152
	stem 4-sided; fls. in whorls; stamens 2 or 4	fls. lilac or white	LABIATAE	166
	stamens 5; fls. white to pink	fls. in small clusters; corolla w. long tube	GENTIANACEAE	153

		Family	Page
	fls. white, in large inflor.; or pink in pairs	CAPRIFOLIACEAE	176
	stamens 4; corolla 5-lobed	VERBENACEAE	165
	stamens alternate w. petals [a]	GENTIANACEAE	153
	stamens opposite petals [b]	PRIMULACEAE	150
Climbing, scrambling or creeping plant	tiny creeping plant; fls. under 3 mm, lvs. rounded	SCROPHULARIACEAE	159
	fls. yellow/green; lvs. palmately lobed	CUCURBITACEAE	130
	fls. white to pink	CONVOLVULACEAE	157
	fls. purple	SOLANACEAE	158
	fls. blue	CAMPANULACEAE	173
Lvs. pinnate		POLEMONIACEAE	154
Lvs. all radical		PLUMBAGINACEAE	149
Lvs. alternate	most stamens obviously hairy	SCROPHULARIACEAE	159
	4 nutlets to be seen inside calyx [c]	BORAGINACEAE	154

▶ 49

long style w. 2 or 3 stigmas	CAMPANULACEAE	173
stem lvs. simple, stalkless; corolla up to 3 mm, white or pink	PRIMULACEAE	150
corolla usu. 5-lobed, purple, white or cream	SOLANACEAE	158

Part O Plants with white or coloured actinomorphic (radially symmetrical) flowers. Sepals and free petals both present.

Inflor. of many fls. florets, or petals in a compact head	See PART K	41
Fls. like tiny Dandelions w. 5–12 petals (florets)	COMPOSITAE Section 3	185
Fls. like whitish or pinkish Daisies, w. 5 broad petals (florets)	lvs. compound; stem woolly	COMPOSITAE	179
Fls. in umbels	petals 5	10 stamens, 5 w. anther & 5 (often bract-like) without	GERANIACEAE	98
		5 stamens only	UMBELLIFERAE	125
	petals 3		ALISMATACEAE	195
Lvs. covered w. long red glands	bog plants	DROSERACEAE	119
Sepals 2	stamens more than 12	PAPAVERACEAE	73
	stamens 2	ONAGRACEAE	120
	stamens 3 or 5	PORTULACACEAE	92

Sepals & petals 3 each	stamens 6 or more; waterside plants	ELATINACEAE	87
	lvs. radical, linear, 3-angled	BUTOMACEAE	196
	lvs. w. a flat blade	ALISMATACEAE	195
	stamens 3; lvs. in pairs	CRASSULACEAE	115
	tiny plant of bare ground	CRASSULACEAE	115
Over 15 stamens	5 fertile and many barren stamens	PARNASSIACEAE	118
	only 1 stem lf., the rest radical		
	lvs. opposite	CISTACEAE	86
	2 small and 3 large sepals		
	stamens in 3 or 5 bundles; no stipules	HYPERICACEAE	85
	lvs. w. stipules	ROSACEAE	110
	petals under 5 mm, lobed; fls. in a spike	RESEDACEAE	83
	lvs. w. stipules	ROSACEAE	110
	stamens free to their base		
	stamens branching from a long tube	MALVACEAE	97
	no stipules	RANUNCULACEAE	68
Petals yellow, under 5 mm, often shrunken or missing	lvs. w. 3 three-toothed lflets	ROSACEAE	110
	Northern hills only		
	lvs. variously lobed or linear	RANUNCULACEAE	68

▶

6 stamens (sometimes 4 long & 2 short)	petals 3 or 6	lvs. mostly in a whorl just below fls.	PRIMULACEAE	150
		lvs. radical	AMARYLLIDACEAE	209
		lvs. in pairs	GENTIANACEAE	153
	petals 4		CRUCIFERAE	75
	petals 5		FRANKENIACEAE	87
Petals 5, deeply cleft, looking like 10			CARYOPHYLLACEAE	87
Stamens twice as many as petals *Note:* some stamens may not have anthers	lvs. succulent, not lobed	plant w/out green lvs.	MONOTROPACEAE	148
		stigmas 2; lvs. edged w. bristles	SAXIFRAGACEAE	117
		stigmas 3	CARYOPHYLLACEAE	87
		stigmas 4 or 5	CRASSULACEAE	115
	lvs. 4 in a whorl; sepals & petals 4 each		TRILLIACEAE	205
	lvs. w. 3 lflets	petals white; stigmas 2; lf. lobes narrow	SAXIFRAGACEAE	117
		lflets broad, heart-shaped	OXALIDACEAE	100
	lvs. lobed, or divided, w. stipules	fls. purplish or purple-veined	GERANIACEAE	98

		fls. yellow or white	ROSACEAE	110
lvs. opposite or whorled		stigmas 2; fr. 2-lobed; lvs. edged w. bristles	SAXIFRAGACEAE	117
		fr. length over 10 times width; lvs usu. lightly toothed [a]	ONAGRACEAE	120
		stamens in 3 bundles; fls. yellow	HYPERICACEAE	85
		fr. a short capsule	CARYOPHYLLACEAE	87
lvs. radical or alternate		fr. over 25 mm long [a]	ONAGRACEAE	120
		petals bent back; fr. a berry	ERICACEAE	146
		fls. white in a simple spike	PYROLACEAE	148
		lvs. palmately lobed	GERANIACEAE	98
		inflor. various	SAXIFRAGACEAE	117
Climbing plant	fls. yellow/green		CUCURBITACEAE	130
No stamens			CARYOPHYLLACEAE	87
Lvs. round, attached at centre to stalk	fls. up to 5 mm across, in tiny umbels	creeping marsh plant	UMBELLIFERAE	125

53

▶

Plant of wet places, over 60 cm high	fls. purple, showy		LYTHRACEAE	119
	fls. yellow		PRIMULACEAE	150
Lvs. linear lanceolate in 1 or more whorls	5 styles		CARYOPHYLLACEAE	87
	5–9 sepals & petals		PRIMULACEAE	150
Lvs. opposite	fls. yellow, red, or pink	lvs. up to 25 mm, linear w. stipules	CARYOPHYLLACEAE	87
		4 or 5 sepals & petals	PRIMULACEAE	150
		6–8 sepals & petals	GENTIANACEAE	153
	fls. blue; lvs. ovate		PRIMULACEAE	150
	lvs. hairy		CARYOPHYLLACEAE	87
	each sepal 3-toothed	sepals & petals 4	LINACEAE	97
	petals shorter or hardly longer than petals		CARYOPHYLLACEAE	87
	petals 5, longer than sepals		LINACEAE	97
Lvs. alternate and/or radical	fls. w. 5 fertile and many barren stamens		PARNASSIACEAE	118
	fls. in a compact inflor. pink or purple		PLUMBAGINACEAE	149

lvs. pinnate	stamens 2, 4 or 6		CRUCIFERAE	75
	stamens 10 or more		ROSACEAE	110
fls. blue; lvs. linear			LINACEAE	97
sepals & petals 5			PRIMULACEAE	150
sepals & petals 4	stamens 6		CRUCIFERAE	75
	stamens 4		PRIMULACEAE	150

Key to species

LYCOPODIACEAE and SELAGINELLACEAE Clubmosses

Plants are often like a stiff robust moss.
Leaves numerous, small, usually overlapping [a].
Spore capsules in terminal cones [b] or at leaf bases [c].

Stems usu. erect, not creeping	lvs. toothed [d]		*Selaginella selaginoides* Lesser Clubmoss
	lvs. not toothed [e]		*Huperzia selago* Fir Clubmoss
lvs. w. hair points [f]	spore cones long-stalked		*Lycopodium clavatum* Stag's-horn Clubmoss
lvs. pressed to stem in 4 ranks			*Diphasiastrum alpinum* Alpine Clubmoss
Lower lvs. not overlapping, ovate	garden escape		*Selaginella kraussiana* Gardeners' Clubmoss
lvs. & cone-scales usu. toothed	Scotland & N. England		*Lycopodium annotinum** Interrupted Clubmoss
stems appear constricted at intervals			
l.f. margins entire	lowland heaths		*Lycopodiella inundata** Marsh Clubmoss

ISOETACEAE Quillwort

An aquatic tufted plant of still water with tubular leaves.

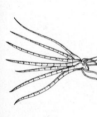

Lvs. stiff, 8–20 cm	in N. & W.	*Isoetes Lacustris** Quillwort

EQUISETACEAE Horsetails

Plants without true leaves, having jointed, sometimes grooved, partly hollow stems, and sometimes also whorls of similar more slender stems. The spores are borne in brownish terminal cones, which in some species appear earlier in the year than the green stems.
Do not confuse Horsetails with the Mare's-tail on p. 000.

Key to barren and fruiting plants

head in fruit

branches
sheath
branch internode
stem
stem cross-section

Fr. stems brownish; green stems absent	sheaths w. 20 or more teeth	spore cone 4–8 cm long	*Equisetum telmateia* Great Horsetail
	sheaths w. 6–12 teeth		*Equisetum arvense* Field Horsetail
	sheaths w. 3–6 teeth	green branches usu. appearing	*Equisetum sylvaticum* Wood Horsetail
Stem w'out whorls of branches	central hollow ⅓ of stem diam.	spore cone 5–7 mm	*Equisetum variegatum** Variegated Horsetail
	central hollow ⅘ of stem diam.	spore 10–20 mm	*Equisetum fluviatile* Water Horsetail

▶

57

Stem branches again branched — *Equisetum sylvaticum* Wood Horsetail

Stems v. finely or hardly ribbed — stems dirty white; branches numerous — *Equisetum telmateia* Great Horsetail

stems green, smooth, 4/5 hollow — usu. in mud or water — *Equisetum fluviatile* Water horsetail

Lowest internode of branches shorter than stem sheath — stem w. 4–8 grooves — *Equisetum palustre* Marsh Horsetail

Lowest internode of branches longer than stem sheath — stem w. 6–18 grooves — *Equisetum arvense* Field Horsetail

OSMUNDACEAE   Royal Fern Family

A large fern whose spores are clustered in a loose spike rising above the leaves.

Lvs. ×2 pinnate, pale green	usu. 50–200 cm high	in marshy/woody places	*Osmunda regalis** Royal Fern

HYMENOPHYLLACEAE   Filmy-ferns

Small fern-like plants with thin translucent leaves.
The spore cases are in a small pouch (indusium) among the leaves.

leaflet

Lflets slightly bent back; indusium entire	mainly in N. and W.	*Hymenophyllum wilsonii* Wilson's Filmy-fern
Indusium toothed	in W. and also The Weald	*Hymenophyllum tunbrigense* Tunbridge Filmy-fern

FERN FAMILIES

ADIANTACEAE, ASPIDIACEAE, ASPLENIACEAE, ATHYRIACEAE, BLECHNACEAE, CRYPTOGRAMMACEAE, HYPOLEPIDACEAE, POLYPODIACEAE, THELYPTERIDACEAE.

The great majority of ferns bear their spores on the backs of the fronds. It is essential to examine a not too ripe reproductive frond in order to identify some of the more difficult species.

To aid identification all the ferns have been included in the key on page 61, though a few of them will be found in other families and keys as well.

There is a simplified key to 18 common ferns on page 252.

Special terms used in the fern key on page 61

Frond   a whole fern leaf (and all its leaflets) rising from the root.
Pinna   a leaflet springing from the main stalk. it may be subdivided into smaller leaflets called pinnules.
Pinnule   a subdivision, or lobe, of a pinna.
Sorus   a patch of spore cases, usually on the back of the frond.
Indusium the cover over a sorus (best seen when young before it shrivels or falls off).

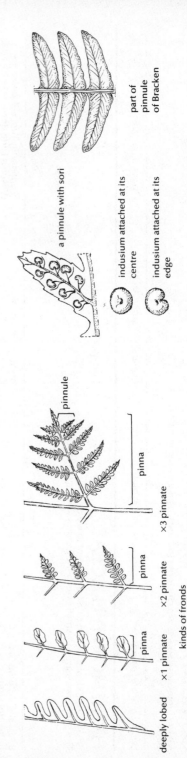

part of
pinnule
of Bracken

a pinnule with sori

indusium attached at its centre

indusium attached at its edge

pinnule

pinna

×3 pinnate

pinna

×2 pinnate

pinna

×1 pinnate

deeply lobed

kinds of fronds

A free-floating aquatic		see page 65	*Azolla filiculoides* Water Fern
Frond simple	Fronds strap-shaped w. sori in lines [a]		*Phyllitis scolopendrium* Hart's-tongue
	usu. only one frond [b]; spore cases in a spike	see page 65	*Ophioglossum vulgatum* Adder's-tongue
Frond pinnate or deeply lobed	pinnae fan-shaped; spore cases on a spike	see page 65	*Botrychium lunaria* Moonwort
	pinnae sharply toothed [c]	mt. plant	*Polystichum lonchitis** Holly Fern
	pinnae entire	sori approx. round [d]	*Polypodium vulgare* Polypody
		sori only on fronds w. v. narrow lobes [e]	*Blechnum spicant* Hard Fern
		fronds covered beneath w. brown scales [f]	*Ceterach officinarum* Rustyback
	stalk green	N. or W. especially on limestone	*Asplenium viride* Green Spleenwort
	stalk almost black	common wall plant [g]	*Asplenium trichomanes* Common Spleenwort
	stalk brown; fronds tough	on sea cliffs [h]	*Asplenium marinum* Sea Spleenwort

Sori along inrolled edge of pinnae see page 60	a	pinnae fan-shaped, on fine stalks	*Adiantum capillus-veneris* Maidenhair Fern
		in tufts up to 15 cm high	N. & W. only	*Cryptogramma crispa* Parsley Fern
		30–200 cm high; fronds arise singly, not in tufts [a]		*Pteridium aquilinum* Bracken
Sori almost cover back of pinnae	b	If. edges rolled back over sori [a]	N. & W. only	*Cryptogramma crispa* Parsley Fern
		lowest pinna the longest	usu. under 12 cm, dull green; pinnae fan-shaped [b]	*Asplenium ruta-muraria* Wall-rue
			10–50 cm; bright green	*Asplenium adiantum-nigrum* Black Spleenwort
	c	lowest pinna not the longest	near the sea	*Asplenium billotii* Lanceolate Spleenwort
Sori in spikes above the fronds			see page 59	*Osmunda regalis** Royal Fern
		Under 10 cm high; pinnae translucent; indusium a 2-lipped capsule	pinnae slightly bent back; indusium entire	see page 59	*Hymenophyllum wilsonii* Wilson's Filmy-fern

indusium toothed		see page 59	*Hymenophyllum tunbrigense* Tunbridge Filmy-fern
Sori approx. oblong or curved, never round a b c sori, with indusia	inner edge of indusium curved [a]	30–100 cm high	*Athyrium filix-femina* Lady-fern
	lowest pinna the longest	usu. under 12 cm, dull green; pinnae fan-shaped; sorus shape [b]	*Asplenium ruta-muraria* Wall-rue
		frond 10–50 cm, bright green; sorus [c]	*Asplenium adiantum-nigrum* Black Spleenwort
	lowest pinna not the longest	near the sea	*Asplenium billotii** Lanceolate Spleenwort
Fronds solitary, not in tufts	fronds mostly 3 times pinnate	pinnules w. pale margin on back; v. common	*Pteridium aquilinum* Bracken
		frond and stalk glandular; on limestone	*Gymnocarpium robertianum** Limestone Fern
		frond glabrous	*Gymnocarpium dryopteris* Oak Fern
	lowest pinnae bent downwards	no indusium	*Phegopteris connectilis* Beech Fern
	marsh plant up to 120 cm		*Thelypteris thelypteroides* Marsh Fern
Indusium under 1 mm or absent	pinnules hardly toothed	sori form neat border round pinnules	*Oreopteris limbosperma* Lemon-scented Fern
	pinnules well toothed	Scottish mts. only	*Athyrium distentifolium** Alpine Lady-fern

63

▶

Indusium attached at its centre; pinnules sharply toothed	frond rigid; about 15 pinnules on longest pinna		*Polystichum aculeatum* Hard Shield-fern
	frond soft; up to 20 pinnules on a pinna		*Polystichum setiferum* Soft Shield-fern
Indusium ovate, pointed	frond delicate, up to 40 cm		*Cystopteris fragilis* Brittle Bladder-fern
Frond ×3 pinnate, or almost so	lower stalk scales w. dark stripe		*Dryopteris dilatata* Broad Buckler-fern
	indusium w. many glands round edge [a]	lowest pinnule on lowest pinna curved	*Dryopteris aemula** Hay-scented Buckler-fern
	indusium entire w'out glands [b]		*Dryopteris carthusiana* Narrow Buckler-fern
Delicate fern up to 40 cm	indusium ovate, pointed, whitish		*Cystopteris fragilis* Brittle Bladder-fern
Sori in neat border round pinnules			*Oreopteris limbosperma* Lemon-scented Fern
Frond fragrant, w. many glands	indusium also glandular	on limestone in N.	*Dryopteris villarii** Rigid Buckler-fern
Main stem v. scaly	pinnae stalks w. blackish patch at base on back		*Dryopteris affinis* Scaly Male-fern
Indusium kidney-shaped		common	*Dryopteris filix-mas* Male-fern

MARSILEACEAE Pillwort Family

A creeping aquatic with fine, cylindrical leaves, which are coiled when young.

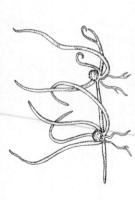

a

Lvs. usu. 3–15 cm [a]	spores in pill-like clusters	in or by acid water

*Pilularia globulifera**
Pillwort

AZOLLACEAE Water Ferns

Small floating plants, often growing in a dense mass, rising above the surface, with hanging rootlets.

b

Lvs. about 1 mm, overlapping [b]	often turns red in Autumn	in still water

*Azolla filiculoides**
Water Fern

OPHIOGLOSSACEAE Adder's-tongue Family

Small ferns with a single leaf up to 20 cm and a reproductive spike rising above it.

c

d

Lf. ovate, entire [c]	in dry grassy places or rock ledges

Ophioglossum vulgatum
Adder's-tongue

Lf. pinnate; lflets fan-shaped [d]	usu. in damp grassy places

Botrychium lunaria
Moonwort

CONIFERAE

As well as the native species of conifers those commonly planted for forestry are included. The latter are marked †.
Except for the Larches all are evergreen.

PINACEAE	CUPRESSACEAE	TAXACEAE
Abies	Chamaecyparis	Taxus
Larix	Juniperus	
Picea	Thuja	
Pinus		
Pseudotsuga		
Tsuga		

a

b

c

Lvs. under 5 mm, scale-like	seaside shrub; fls. small, pink, in spikes	(not a conifer)	TAMARICACEAE page 86
	topmost shoot droops; lvs. parsley-scented	cone scales do not overlap [b]	*Chamaecyparis lawsoniana*† Lawson's Cypress
	lvs. pineapple-scented	cone scales overlap [c]	*Thuja plicata*† Western Red Cedar
Lvs. over 8 cm, in bundles of 3			*Pinus radiata*† Monterey Pine
Lvs. about 1 cm, in whorls of 3 [a]	shrub; fr. purple, juicy		*Juniperus communis* Juniper
Lvs. in bundles of 2	lvs. mostly under 10 cm	lvs. bluish; upper bark orange	*Pinus sylvestris* Scots Pine
		lvs. dark green; on moors in N. & W.	*Pinus contorta*† Lodgepole Pine
	lvs. 10–20 cm	cone shiny, 10–15 cm; in S. only	*Pinus pinaster*† Maritime Pine
		cone 5–10 cm	*Pinus nigra*† Corsican Pine

Most lvs. in clusters of 20 or so on v. short stalks; deciduous	tips of cone scales upright [a] (a)	lvs. bright green	*Larix decidua*† European Larch
	tips of cone scales curl out & down [b] (b)	lvs. blue-green	*Larix kaempferi*† Japanese Larch
	cone scales curl out but not down		*Larix × eurolepis*† Hybrid Larch
Woody lf. bases remain like pegs after lvs. fall [c] (c)	lvs. 4-sided, all green	cone 12–15 cm	*Picea abies*† Norway Spruce
	lvs. flat, keeled, stiff, dark above, pale below	cone 5–8 cm	*Picea sitchensis*† Sitka Spruce
Lf. scars round, flat; cones erect	lf. sprays flattish		*Abies grandis*† Giant Fir
	lvs. curved, bluish		*Abies procera*† Noble Fir
Lf. scars like bumps	topmost shoot droops	lvs. of various lengths; cone 2–3 cm	*Tsuga heterophylla*† Western Hemlock
	cones w. 3-forked bracts [d] cones 5–8 cm	(d)	*Pseudotsuga menziesii*† Douglas Fir
Lvs. dark green	seed surrounded by red fleshy cup when ripe		*Taxus baccata* Yew

RANUNCULACEAE Buttercup family

Plants with a variety of form and colour. The flowers have more than 12 stamens (except *Myosurus*) and the fruit is often composed of a number of distinct parts.
Sepals and/or petals are often in 5's.

Leaves are alternate (except *Clematis*), often lobed, and do not bear stipules.
Similar-looking flowers (i.e. with many stamens) may belong to the
Rose family – on p. 110
St John's Wort family – on p. 85.

Woody climber w. creamy flowers	only 4 sepals (which look like petals)	lvs. opposite, pinnate, lflets. well spaced	*Clematis vitalba* Traveller's-joy
Ea. fl. w. 5 tubular petals & 5 spurs			*Aquilegia vulgaris* Columbine
Fls. scarlet; lvs. finely dissected	calcareous soil		*Adonis annua** Pheasant's-eye
Fls. blue/violet/purple	fls. zygomorphic w. large hood	S.W. (except for introductions)	*Aconitum napellus** Monk's-hood
	perianth w. 6 segments	in calcareous turf	*Pulsatilla vulgaris** Pasqueflower
Lvs. linear in a rosette 	petals up to 5 mm, greenish yellow; fr. spike up to 7 cm	up to 12 cm tall; may look like a Plantain	*Myosurus minimus* Mousetail
Stamens longer than the 4 sepals; lvs. twice or more pinnate  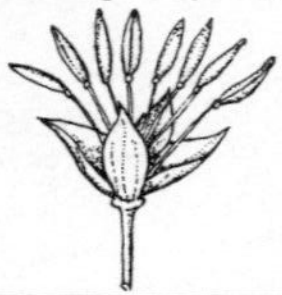	up to 15 cm tall; fls. in simple loose spike	mts. in N.	*Thalictrum alpinum* Alpine Meadow-rue
	stamens mostly erect		*Thalictrum flavum* Common Meadow-rue
	stamens drooping		*Thalictrum minus* Lesser Meadow-rue

Fls. 1–5 cm across, greenish	upper lvs. simple; sepals usu. w. purple border	calcareous woods	*Helleborus foetidus* * Stinking Hellebore
	upper lvs. lobed	calcareous woods	*Helleborus viridis* Green Hellebore
Water or waterside plants w. white petals NOTE: some of these plants vary a great deal according to water depth and exposure	broad floating lvs. present but no fine submerged lvs.; up to 10 stamens	lvs. lobed over ½ way in; lobes well spaced	*Ranunculus tripartitus* * Three-lobed Crowfoot
		fls. up to 6 mm; lf lobes broadest at base [a]	*Ranunculus hederaceus* Ivy-leaved Crowfoot
		fls. 8–12 mm; lf. lobes narrowed at base [b]	*Ranunculus omiophyllus* Round-leaved Crowfoot
	broad floating & fine submerged lvs. both present	fr. of 40–100 glabrous achenes; near the sea	*Ranunculus baudotii* Brackish Water-crowfoot
		fls. 12–25 mm; stamens 15 or more; achenes usu. hairy	*Ranunculus aquatilis* Common Water-crowfoot
		fls. 3–7 mm; up to 10 stamens	*Ranunculus tripartitus* * Three-lobed Crowfoot
	submerged lvs. w. short rigid lobes [d]	outline of lf. rounded	*Ranunculus circinatus* Fan-leaved Water-crowfoot
	submerged lvs. 10–30 cm long; fls. 20–30 mm [e]	in rivers	*Ranunculus fluitans* River Water-Crowfoot
	submerged lvs. short, tassel-like; fls. 5–10 mm	in still water	*Ranunculus trichophyllus* Thread-leaved Water-crowfoot
	fls. 12–25 mm; stamens 15 or more	often floating and submerged lvs. both present [c]	*Ranunculus aquatilis* Common Water-Crowfoot

floating leaves

a b

submerged leaves

c d

e

Woodland plants w. white to pale purpls fls.	fls. solitary; perianth usu. w. 6 segments	in fl. March–May	*Anemone nemorosa* Wood Anemone
	fls. in clusters	calcareous places in N. England	*Actaea spicata** Baneberry
Lvs. not or only shallowly lobed, or with blunt teeth	lvs. linear/lanceolate margins entire or lightly toothed	fls. 7–18 mm across; plant up to 50 cm high	*Ranunculus flammula* Lesser Spearwort
		fls. 20–50 mm across; plant plant 50–120 cm high	*Ranunculus lingua* Greater Spearwort
	sepals 3; petals usu. 8	lvs. cordate w. blunt teeth	*Ranunculus ficaria* Lesser Celandine
	sepals 5, large; yellow; no petals	lvs. cordate, reniform, w. blunt teeth	*Caltha palustris* Marsh-marigold
A frill of lflets. just below fl. [a]	fls. in early spring	glossy, lobed, radical lvs. appear later	*Eranthis hyemalis** Winter Aconite
Fls. almost spherical, 25–30 mm across		Wales & N.	*Trollius europaeus* Globeflower
Fr. spiny [b]	fls. 4–12 mm diam.	cornfields etc.	*Ranunculus arvensis* Corn Buttercup
Lvs. almost hairless	petals hardly longer than sepals, well spaced	in muddy places	*Ranunculus sceleratus* Celery-leaved Buttercup
	petals overlapping, but often imperfect	woodland	*Ranunculus auricomus* Goldilocks

a   b

Sepals turned down in fl.	petals up to 3 mm, often absent	a low, sprawling, hairy plant	*Ranunculus parviflorus* Small-flowered Buttercup
	stem base bulbous; achenes (1-seeded fr.) smooth	v. common	*Ranunculus bulbosus* Bulbous Buttercup
	achenes (1-seeded fr.) w. tiny warts		*Ranunculus sardous* Hairy Buttercup
Fl. stem lightly grooved			*Ranunculus repens* Creeping Buttercup
Fl. stem not grooved	lvs. palmately lobed, deeply cut	the tallest of the common Buttercups	*Ranunculus acris* Meadow Buttercup

BERBERIDACEAE Barberry Family

Shrubs with spiny twigs or leaves.
The yellow petals are in several whorls, and the flowers in clusters.

a

Branches w. 3-forked spines [a]	fr. red, oblong; lvs. toothed		*Berberis vulgaris** Barberry
Lvs. pinnate, evergreen, Holly-like	fr. blue-black	introduced plant	*Mahonia aquifolium** Oregon-grape

NYMPHAECEAE Water-lilies

Aquatics with large floating leaves. Sepals green or yellow 4-6.
Petals & stamens numerous.

Petals white, about 20	*Nymphaea alba* White Water-lily
Petals yellow	petals 5, larger than sepals	MENYANTHACEAE page 154
	petals many, much smaller than the yellow sepals	*Nuphar lutea* Yellow Water-lily

CERATOPHYLLACEAE Hornworts

Submerged aquatics with fine, forked leaves, forming tassels at the stem
tips.
The tiny flowers are in the leaf axils.
Growing in still, fresh or brackish water.

a

b

c

Fr. w. 2 basal pines [a]	lvs. once or twice forked [b]	*Ceratophyllum demersum* Rigid Hornwort
Fr. w'out spines	lvs. 3 times forked [c]	*Ceratophyllum submersum** Soft Hornwort

PAPAVERACEAE Poppy Family

Poppies have 2 sepals which usually fall when the flower opens; 4 petals and many stamens.
The lobed leaves and stem contain a milky, yellowish or bright orange juice.

   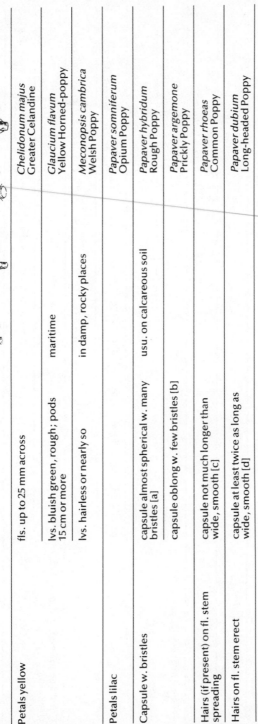

	a	b	c	d	
Petals yellow			fls. up to 25 mm across		*Chelidonum majus* Greater Celandine
	maritime		lvs. bluish green, rough; pods 15 cm or more		*Glaucium flavum* Yellow Horned-poppy
	in damp, rocky places		lvs. hairless or nearly so		*Meconopsis cambrica* Welsh Poppy
Petals lilac					*Papaver somniferum* Opium Poppy
Capsule w. bristles	usu. on calcareous soil		capsule almost spherical w. many bristles [a]		*Papaver hybridum* Rough Poppy
			capsule oblong w. few bristles [b]		*Papaver argemone* Prickly Poppy
Hairs (if present) on fl. stem spreading			capsule not much longer than wide, smooth [c]		*Papaver rhoeas* Common Poppy
Hairs on fl. stem erect			capsule at least twice as long as wide, smooth [d]		*Papaver dubium* Long-headed Poppy

FUMARIACEAE Fumitory Family

Generally sprawling plants with many-lobed leaves.
The zygomorphic flowers have 2 sepals, and stamens in 2 groups of 3.

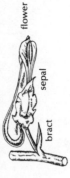

flower / sepal / bract

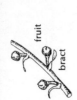

fruit / bract

			Species
Many lvs. end w. a tendril	petals creamy		*Corydalis claviculata* Climbing Corydalis
Fls. bright yellow	usu. on walls		*Corydalis lutea* Yellow Corydalis
Sepals at least as wide as corolla	corolla creamy w. dark tips, up to 12 mm		*Fumaria capreolata* White Ramping-fumitory
	corolla about 6 mm; lf. lobes channelled	Eastern counties	*Fumaria densiflora* Dense-flowered Fumitory
	corolla 9–13 mm; lf. lobes ovate, flat	about 20 fls. to each head	*Fumaria purpurea** Purple Ramping-fumitory
		about 12 fls. to a head (sepals usu. narrower)	*Fumaria muralis* Common Ramping-fumitory
Corolla up to 6 mm; usu. plants of chalky soil	lf. lobes channelled	bracts as long as fr. stalks	*Fumaria parviflora** Fine-leaved Fumitory
	lf. lobes flat	bracts shorter than fr. stalks	*Fumaria vaillantii** Few-flowered Fumitory
Sepal teeth mostly near base	margin of lowest petal w. flat spreading tip	common	*Fumaria officinalis* Common Fumitory
	margin of lowest petal v. narrow to end	frequent in W.	*Fumaria muralis* Common Ramping-fumitory
Sepals toothed all round		rare, except in W.	*Fumaria bastardii** Tall Ramping-fumitory

CRUCIFERAE Cabbage Family

A typical Crucifer flower has 4 sepals, 4 petals and 6 stamens (a, b) though
a few species are missing petals or stamens.
The fruit is a capsule which may be long and narrow (c) or pouch-like (d).
It is often needed for identification.
Leaves are alternate, without stipules.
Flower colours in some species are very variable; many of the usually pale
purple ones range from white to a medium purple. Such plants are
included in two or even three sections of the key.

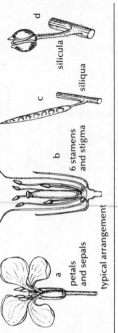

petals
and sepals

a

typical arrangement

b

6 stamens
and stigma

c

siliqua

d

silicula

Petals absent	Section 1 page 75
Petals pale or bright yellow	Section 2 page 76
Petals white	Section 3 page 79
Petals lilac or purple (often pale)	Section 4 page 82

CRUCIFERAE Section 1 Flowers without petals

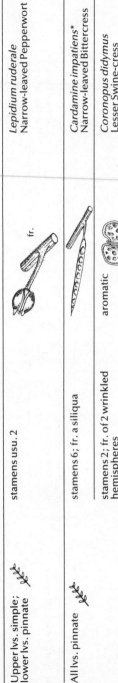

Upper lvs. simple; lower lvs. pinnate	stamens usu. 2	fr.	*Lepidium ruderale* Narrow-leaved Pepperwort
All lvs. pinnate	stamens 6; fr. a siliqua		*Cardamine impatiens* * Narrow-leaved Bittercress
	stamens 2; fr. of 2 wrinkled hemispheres	aromatic	*Coronopus didymus* Lesser Swine-cress

CRUCIFERAE Section 2 Flowers yellow

Lf. margins almost entire	fl. 25 mm across		*Cheiranthus cheiri* Wallflower
	fl. about 6 mm across		*Erysimum cheiranthoides* Treacle Mustard
Smell obnoxious (crush the stem); fls. pale yellow; lvs. pinnately lobed	fr. nearly parallel to stem [a]	stem glabrous	*Diplotaxis tenuifolia* Perennial Wall-rocket
	fr. inclined away from stem [b]	stem slightly hairy; fr. stalk much shorter than fr.	*Diplotaxis muralis* Annual Wall-rocket
Lvs. 3 times pinnate; finely divided			*Descurainia sophia* Flixweed
Fr. jointed like a string of beans	usu. 1–4 seeds in deeply jointed fr.	beak hardly ×2 length of top joint	*Raphanus maritimus* Sea Radish
	usu. 3–8 seeds; fr. w. shallower joints	beak over ×2 length of upper joints	*Raphanus raphanistrum* Wild Radish
Fr. globular or short (length up to 4 times width but often less)	stem & fr. covered w. warts		*Bunias orientalis* Warty Cabbage
	fr. opens longitudinally into 2 halves	fr. ovoid, under half length of stalk [c]	*Rorippa amphibia* Great Yellow-cress
		fr. oblong, about equal to stalk [d]	*Rorippa palustris* Marsh Yellow-cress

	fr. 2-jointed, upper joint larger	lower lvs. w. about 6 prs. lobes; fls. bright	*Rapistrum perenne* Steppe Cabbage
		lower lvs. w. about 3 prs. lobes; fls. pale	*Rapistrum rugosum* Bastard Cabbage
	usu. 1–4 seeds in deeply jointed pod [a]	maritime [a]	*Raphanus maritimus* Sea Radish
Fr. pressed close to stem	fls. about 3 mm across		*Sisymbrium officinale* Hedge Mustard
	upper lvs. clasp stem	upper lf. margins entire	*Arabis glabra** Tower Mustard
		upper lvs. broadly ovate, toothed, w. auricles	*Barbarea vulgaris* Winter-cress
	upper lvs. pinnate	upper lvs. pinnate	*Barbarea intermedia* Medium-flowered Winter-cress
			Barbarea intermedia Medium-flowered Winter-cress
	lower parts of plant densely hairy	usu. one seed in beak of fr.	*Hirschfeldia incana** Hoary Mustard
	lower part of plant merely bristly	no seeds in the beak	*Brassica nigra* Black Mustard
Fr. bristly	upper lvs. lanceolate, but not lobed	fr. beak conical	*Sinapis arvensis* Charlock
	all lvs. lobed	fr. beak flattened	*Sinapis alba* White Mustard
Cross-section of ripe fr. squarish; beak under 3 mm	lower lvs. pinnate	upper lvs. broad, toothed, not pinnate	*Barbarea vulgaris* Winter-cress
		lower lvs. w. 6–10 prs. lobes; fr. over 3 cm	*Barbarea verna* American Winter-cress

▶

Fruit character	Leaf / plant character	Further character	Species
Fr. ends w. a narrow beak	lvs. lanceolate w. a few teeth	If. lobes fewer; fr. shorter	*Barbarea intermedia* Medium-flowered Winter-cress
			Erysimum cheiranthoides Treacle Mustard
	only one vein on each fr.-half	all lvs. hairless; open fls. well below buds	*Brassica oleracea** Wild Cabbage
		open fls. higher than buds, bright yellow	*Brassica rapa* Wild Turnip
		fls. pale yellow	*Brassica napus* Rape
	on lower lvs. terminal lobe much the largest		*Sinapis arvensis* Charlock
	plant almost hairless; few stem lvs.	W. coast	*Rhynchosinapis monensis** Isle of Man Cabbage
	stem & lvs. hairy below	a casual	*Rhynchosinapis cheiranthos* Wallflower Cabbage!
Fr. often over 4 cm long	upper lvs. w. several pairs of v. narrow lobes		*Sisymbrium altissimum* Tall Rocket
	If. lobes few; end lobe lanceolate		*Sisymbrium orientale* Eastern Rocket
Fr. under 4 cm	stem & lvs. usu. w. stiff hairs	fr. 25–40 mm, often hairy	*Sinapis arvensis* Charlock
	plant hairless	fr. under 20 mm	*Rorippa sylvestris* Creeping Yellow-cress

CRUCIFERAE Section 3 Flowers white

Lvs. large, thick, bluish, cabbage-like	maritime	fr. c. 12 mm	*Crambe maritima* Sea-kale
Larger lvs. dock-like, sometimes lobed	fls. 8 mm across; usu. some lvs. lobed	fr. c. 5 mm	*Armoracia rusticana* Horse-radish
	fls. 3 mm across; lvs. simple	fr. c. 2 mm	*Lepidium latifolium!* Dittander
Lower lvs. large, rounded, cordate	lvs. smell of garlic	fr. 35–60 mm	*Alliaria petiolata* Garlic Mustard
Most lvs. pinnate	lvs. often succulent; fr. of 2 unequal segments [h]	in maritime sand or shingle	*Cakile maritima* Sea Rocket
	fr. of 2 wrinkled hemispheres	petals shorter than sepals; stamens 2; fr. [a]	*Coronopus didymus* Lesser Swine-cress
		petals longer than sepals; stamens 6; fr. [b]	*Coronopus squamatus* Swine-cress
	aquatic	seeds in 2 rows [c]	*Nasturtium officinale* Water-cress
		seeds in 1 row [d]; lvs. turn brown in Autumn	*Nasturtium microphyllum* Brown-leaved Water-cress
	fr. a small pouch (up to 6 mm)	upper lvs. nearly simple; fr [e]	*Lepidium sativum** Garden Cress
		stem bears only a few (toothed) lvs.; fr [f]	*Teesdalia nudicaulis* Shepherd's Cress

h

a b
c d

e f

▶

79

		a	some pinnate lvs. on stem; fr. [a]	*Hornungia petraea** Hutchinsia
		upper lvs. simple, toothed	fr.	*Cardaminopsis petraea** Northern Rock-cress
		fls. under 1 cm	lf. segments narrow, sharply toothed	*Cardamine impatiens** Narrow-leaved Bitter-cress
			only 4 stamens; young fr. often overtop fl. buds	*Cardamine hirsuta* Hairy Bitter-cress
			6 stamens; young fr. usu. not overtopping fl. buds	*Cardamine flexuosa* Wavy Bitter-cress
		fls. over 1 cm; in damp places	anthers yellow	*Cardamine pratensis* Cuckooflower
			anthers violet	*Cardamine amara* Large Bitter-cress
	Tiny aquatic up to 8 cm w. linear lvs.	in N. & W.	fr.	*Subularia aquatica** Awlwort
	Fr. jointed like a string of beans			*Raphanus raphanistrum* Wild Radish
	Fr. a triangular or heart-shaped pouch	basal lvs. almost entire; fr. winged [b]	uncommon, on limestone	*Thlaspi alpestre* Alpine Penny-cress
b		lvs. variously lobed or toothed; fr. [c] c	v. common	*Capsella bursa-pastoris* Shepherd's-purse
	Usu. tiny plant up to 20 cm w. no stem lvs.		dry places	*Erophila verna* (agg.) Whitlowgrass
			fr.	

Fr. a cylindrical siliqua (see p. 00)	fls. 18 mm or more across, white or purple	lf. margins entire [a] on sea cliffs in S.		*Matthiola incana** Hoary Stock
		lvs. toothed [b]		*Hesperis matronalis* Dame's-violet
	lower lvs. deeply lobed to pinnate	upper lvs. w. auricles [c]		*Arabis glabra** Tower Mustard
		no auricles to upper lvs [d]		*Cardaminopsis petraea*! Northern Rock-cress
	pods erect, close to stem	all lvs. hairy [e]		*Arabis hirsuta* Hairy Rock-cress
		upper lvs. almost hairless w. auricles [c]		*Arabis glabra** Tower Mustard
	upper lvs. toothed clasping stem	fls. about 15 mm across		*Arabis caucasica** Garden Arabis
	upper lvs. entire [f]	fls. about 3 mm across		*Arabidopsis thaliana* Thale Cress
All lvs. linear, entire				*Lobularia maritima* Sweet Alison
Only 1 seed in each pod-half	outer petals much larger than inner ones	Chilterns; fr. [g]		*Iberis amara** Wild Candytuft
	fls. about 3 mm across; fr. winged at top	style longer than wing tips; anthers violet; fr. [h]		*Lepidium heterophyllum* Smith's Pepperwort
		style much shorter; anthers yellow; fr. [i]		*Lepidium campestre* Field Pepperwort
	fls. 5–6 mm across; fr. not winged	fr [j]		*Cardaria draba* Hoary Cress

fr.

a b c d e f g h i j

▶

Radical lvs. w. long stalks		
upper lvs. ivy-shaped w. short stalks [a]	maritime	*Cochlearia danica* Danish Scurvygrass
radical lvs. w. cordate bases [b]		*Cochlearia officinalis* Common Scurvygrass
radical lvs. w. tapering bases [c]	in saltmarshes	*Cochlearia anglica* English Scurvygrass
Fr. flat, round, w. a notched wing		*Thlaspi arvense* Field Penny-cress
Fr. twisted	in North	*Draba incana** Hoary Whitlowgrass
Fr. straight		*Draba muralis** Wall Whitlowgrass

CRUCIFERAE Section 4 Flowers pale lilac to purple

Fr. jointed like a string of beads [d]	petals purple-veined	*Raphanus raphanistrum* Wild Radish
Lvs. long, entire, downy	on sea cliffs in S.	*Matthiola incana** Hoary Stock
Upper lvs. stalked, ivy-shaped [e]	maritime	*Cochlearia danica* Danish Scurvygrass

Lvs. pinnate	lvs. often succulent, in maritime sand or shingle	fr.	*Cakile maritima* Sea Rocket
	upper lvs. w. axillary bulbils; no basal rosette of lvs.	usu. in calcareous woods	*Cardamine bulbifera** Coralroot
	no bulbils in upper lvs.; rosette or lvs. at base		*Cardamine pratensis* Cuckooflower
	small alpine, up to 25 cm; fls. about 6 mm across	upper lvs. lanceolate	*Cardaminopsis petraea** Northern Rock-cress
Radical lvs. long-stalked, margins entire	on limestone hills		*Thlaspi alpestre** Alpine Penny-cress
Fls. about 18 mm across; plant 40–80 cm high			*Hesperis matronalis* Dame's-violet
Uncommon alpine up to 25 cm; fls. about 6 mm	lower lvs. almost pinnate, upper lanceolate		*Cardaminopsis petraea** Northern Rock-cress

RESEDACEAE Mignonettes

The Mignonettes grow from 30 to 150 cm high and bear white or yellow flowers in long spikes. The 4 to 6 petals are lobed, and the stamens numerous and prominent.
The two yellow-flowered species grow typically in disturbed ground, especially on chalk or limestone.

lobed petal

many stamens

Lvs. deeply lobed	petals greenish-yellow, usu. 6	capsule 3-lobed	*Reseda lutea* Wild Mignonette
	petals white, usu. 5	capsule 4-lobed	*Reseda alba* White Mignonette
l.f. margins entire w. wavy edges	petals greenish-yellow, usu. 4	capsule 3-lobed	*Reseda luteola* Weld

83

VIOLACEAE Violets and Pansies

Viola flowers are zygomorphic with 5 sepals and 5 petals, the lowest bearing a lip in front and a spur behind. At the base of the leaf stalks are stipules which may be simple, toothed or lobed. Hybrids are frequent in this family.

Sepals obtuse/rounded	lvs. almost round; stalks hairless	marsh plant	*Viola palustris* Marsh Violet
	lf. stalks w. spreading hairs and bracts at or above half way	scented; fls. usu. deep purple or white; common	*Viola odorata* Sweet Violet
	lf. stalks w. hairs bent down and bracts below half way	not scented; chiefly in calcareous turf	*Viola hirta* Hairy Violet
Stipules pinnate.; lower petal usu. creamy/yellow w. violet markings, or mostly violet	fl. under 15 mm across vertically	petals shorter than sepals	*Viola arvensis* Field Pansy
	fl. over 20 mm across vertically	petals longer than sepals; hills in N.	*Viola lutea* Mountain Pansy
	intermediate in size between last 2 species	petals longer than sepals	*Viola tricolor* Wild Pansy
Lf. base cuneate	on heaths		*Viola lactea* Pale Dog-violet
	petals bluish; stipules w. short teeth		*Viola canina* Heath Dog-violet
Corolla spur paler than petals	petals violet; stipules w. long fine teeth		*Viola riviniana* Common Dog-violet
Spur dark			*Viola reichenbachiana* Early Dog-violet

POLYGALACEAE Milkworts

Dainty plants of grassland and heathland with narrowly lanceolate leaves. The 6–8 mm zygomorphic flowers are most often blue but vary from white to pinkish purple.
There are 5 sepals, but 2 are much larger than the others, coloured, and easily mistaken for petals. The 3 petals and 8 stamens all spring from an inner tube.
Flowers blue, pink or white.

Most lower lvs. longer than upper ones	veins on larger sepals hardly branched	in chalk grassland	*Polygala calcarea* Chalk Milkwort
Lower lvs. alternate	veins on larger sepals well branched		*Polygala vulgaris* Common Milkwort
Lower lvs. opposite			*Polygala serpyllifolia* Heath Milkwort

GUTTIFERAE or HYPERICACEAE St John's-wort

All St John's-Worts have untoothed opposite leaves and yellowish flowers with 5 sepals, 5 petals, and numerous stamens arranged in either 3 or 5 bundles.

Fls. 7–8 cm across		a garden escape	*Hypericum calycinum* Rose-of-Sharon
Small shrub	fr. a berry – red turning black	in shady places	*Hypericum androsaemum* Tutsan
	creeping bog plant		*Hypericum elodes* Marsh St John's-wort
	erect plant		*Hypericum hirsutum* Hairy St John's-wort
Trailing plant w. v. slender stems	fls. about 1 cm across; sepals unequal		*Hypericum humifusum* Trailing St John's-wort
Lvs. slightly downy beneath	fls. in a close cluster	usu. on calcareous soil	*Hypericum montanum* Pale St John's-wort

▶

Sepals edged w. black glands	lvs. edged w. black dots beneath	usu. on calcareous soil	*Hypericum montanum* Pale St John's-wort
	lvs. dotted w. translucent spots	stem w. 2 faint ridges	*Hypericum perforatum* Perforate St John's-wort
		no ridges on stem; petals tinged w. red	*Hypericum pulchrum* Slender St John's-wort
Stem w. 4 ridges	lf. margins wavy	in S.W.	*Hypericum undulatum** Wavy St John's-wort
	stem wings prominent; sepals lanceolate acute		*Hypericum tetrapterum* Squate-stalked St John's-wort
	sepals ovate	lvs. w. translucent dots	*Hypericum maculatum* Imperforate St John's-wort
Stems w. 2 raised lines	sepals acute	few or no dots on lvs.	*Hypericum perforatum* Perforate St John's-wort

CISTACEAE Rock-roses

An undershrub with opposite leaves and stipules.
The flower has 2 small sepals, 3 large sepals, 5 petals and numerous stamens.

Fls. yellow; about 2 cm across; stems woody	lvs. elliptical, whitish beneath	often in calcareous grassland	*Helianthemum nummularium* Common Rock-rose

TAMARICEAE Tamarisks

A planted and naturalized seaside shrub growing to about 3 m.
The feathery foliage is composed of minute overlapping leaves.

Lvs. about 2 mm	fls. pink in spikes; sepals, petals, stamens all 5	*Tamarix gallica* Tamarisk

FRANKENIACEAE Sea-heath Family

A low Heather-like undershrub growing on the drier parts of saltmarshes.
The leaves may be opposite or whorled, and have inrolled margins.

Lvs. 2–4 mm opposite or in whorls, linear	petals 5, pink, crinkly; stamens 6	S.E. coasts	*Frankenia laevis** Sea-heath

ELATINACEAE Waterworts

A small, creeping waterside plant, which may be quite submerged, with opposite, entire, spoon-shaped leaves.

Petals pink/white; sepals and petals 3	stamens 6	*Elatine hexandra** Six-stamened Waterwort

CARYOPHYLLACEAE Campion Family

A large family. Leaves are opposite, or occasionally in clusters apparently whorled.
Sepals and petals are normally in 4's or 5's, with usually twice as many stamens. There may be 2, 3 or 5 styles.
There are no blue or yellow flowers in this family.

Petals under 3 mm or absent	sepals usu. 5	small, compact yellow-green alpine; 3 styles; Scot.	*Minuartia sedoides* Cyphel
		fls. in small clusters; lvs. linear; dry places [a]	*Scleranthus annuus* Annual Knawel
		lvs. ovate, stalked; stamens 1–3	*Stellaria pallida* Lesser Chickweed

▶

87

Lvs. whorled, in clusters of 8 or more	lvs. end w. minute bristle	central rosette w'out fls.; perennial	*Sagina procumbens* Procumbent Pearlwort
		fls. from centre [a]; annual	*Sagina apetala* Annual Pearlwort
	lvs. w'out bristle at end		*Sagina maritima* Sea Pearlwort
Small plants usu. well under 30 cm w. narrow lvs. in whorls [b] or pairs, a tiny stipule at their base; petals usu. pink-tinged	fls. white; styles 5		*Spergula arvensis* Corn Spurrey
	whole plant densely covered w. glands	seaside cliffs & walls	*Spergularia rupicola* Rock Sea-spurrey
	petals at least as long as sepals; 10 stamens	all seeds winged [c]; saltmarshes etc.	*Spergularia media* Greater Sea-spurrey
	stipules silvery, lanceolate; fls. 3–5 mm across	drier sandy places	*Spergularia rubra* Sand Spurrey
	fls. 6–7 mm across; stamens 8 or fewer	a few seeds winged; saltmarshes	*Spergularia marina* Lesser Sea-spurrey
Sepals longer than petals (often about 1½ times as long)	petals 5, entire; stamens 10; lvs. ovate	lvs. 3-veined, over 10 mm long	*Moehringia trinervia* Three-nerved Sandwort
		capsule straight-sided [d]	*Arenaria leptoclados* Slender Sandwort
		capsule convex at base [e]	*Arenaria serpyllifolia* Thyme-leaved Sandwort
	petals v. deeply cleft [f]		*Stellaria alsine* Bog Stitchwort
	petals, stamens & styles 4 each		*Moenchia erecta* Upright Chickweed
Petals red, each 4-lobed		in wet places	*Lychnis flos-cuculi* Ragged-Robin

Petals pink, spotted, w. toothed edges	fls. 8–12 mm across, usu. in clusters	*Dianthus armeria* Deptford Pink
	fls. about 18 mm, usu. in ones	*Dianthus deltoides** Maiden Pink

calyx

Sepals joined at base into a tube	fls. pink	styles 2	*Saponaria officinalis* Soapwort
		styles 5	
		fls. white	*Silene alba* White Campion
		petals lobed about ½-way, red [a]	*Silene dioica* Red Campion
		petals only notched, red [b]; lvs. narrowly lanceolate	*Lychnis viscaria** Sticky Catchfly
		styles 0	
		mt. plant under 10 cm high; fls. pink; petal [c]	*Silene acaulis* Moss Campion
		fls. 3–4 mm across; petals narrow, entire [d]	*Silene otites** Spanish Catchfly
		stamens protrude well beyond corolla tube; petal [e]	*Silene nutans* Nottingham Catchfly
		fls. white; petal [f]	*Silene alba* White Campion
		fls. red; petal [a]	*Silene dioica* Red Campion

petals
a   b

styles 5

petals
c   d

styles 0

e   f

compact mt. plant under 10 cm high	fls. usu. pink; lvs. edged w. bristles; petal [c] — *Silene acaulis* Moss Campion
lip of petals lobed to ½-way or more	fls. pinkish, opening at night; petal [g] — *Silene noctiflora* Night-flowering Catchfly
	whole plant sticky/hairy; petal [e] — *Silene nutans* Nottingham Catchfly

g

▶

a   b

petal

calyx sticky/hairy; petals slightly notched	fls. mostly in ones or twos; ripe capsule teeth recurved [a + b]	*Silene maritima* Sea Campion
	fls. in loose clusters; capsule teeth usu. erect	*Silene vulgaris* Bladder Campion
	calyx w. 25–30 veins	*Silene conica** Sand Catchfly
	calyx w. 10 veins	*Silene gallica* Small-flowered Catchfly
petals narrow	only in E. Anglia	*Silene otites** Spanish Catchfly
4 sepals, petals & styles		
petals notched	(fl. parts occasionally in 5's)	*Cerastium diffusum* Sea Mouse-ear
petals obtuse		*Moenchia erecta* Upright Chickweed
Petal margins quite entire		
lvs. ovate, succulent, up to 8 seeds in a capsule	on dunes & shingle	*Honkenya peploides* Sea Sandwort
styles 5	petals twice length of sepals	*Sagina nodosa* Knotted Pearlwort
	petals about equal to sepals	*Sagina subulata* Heath Pearlwort
lvs. linear	petals longer than sepals	*Minuartia verna* Spring Sandwort
	petals shorter than sepals	*Minuartia hybrida* Fine-leaved Sandwort
lvs. 3-veined, over 10 mm long		*Moehringia trinervia* Three-nerved Sandwort

	capsule straight-sided		*Arenaria leptoclados* Slender Sandwort
	capsule convex at base		*Arenaria serpyllifolia* Thyme-leaved Sandwort
Petals cleft almost to base	styles 5		*Myosoton aquaticum* Water Chickweed
	lower lvs. stalked	petals twice as long as sepals	*Stellaria nemorum* Wood Stitchwort
		10 stamens	*Stellaria neglecta* Greater Chickweed
		3–8 stamens	*Stellaria media* Common Chickweed
	fls. 18 mm or more across	(petals more usu. lobed about ½-way) v. common	*Stellaria holostea* Greater Stitchwort
	fls. 12 mm or more; petals much longer than sepals	uncommon marsh plant	*Stellaria palustris* Marsh Stitchwort
	fls. smaller; petals not much longer than sepals	common	*Stellaria graminea* Lesser Stitchwort
Styles 3	plant 15–60 cm high; lvs. 4 cm or more	v. common	*Stellaria holostea* Greater Stitchwort
	plant up to 10 cm	mts. in Scotland	*Cerastium cerastoides** Starwort Mouse-ear

Petals nearly twice length of calyx; fls. mostly in ones	If. hairs long, white, dense	on mts.	*Cerastium alpinum* Alpine Mouse-ear
	If. hairs short, white, dense	escaped garden plant	*Cerastium tomentosum* Snow-in-Summer
	plant hairy but not densely so		*Cerastium arvense* Field Mouse-ear
Stamens 10	fls. in compact clusters; lvs. often pale	glands among hairs in upper parts	*Cerastium glomeratum* Sticky Mouse-ear
	fls. in loose clusters		*Cerastium fontanum* Common Mouse-ear
Upper lvs. w. transparent tips or edges; 5 stamens & styles	these lvs. nearly ½ transparent [a]	petals shorter than sepals	*Cerastium semidecandrum* Little Mouse-ear
	only margins of these lvs. transparent	petals about equal to sepals; calcareous places in S.	*Cerastium pumilum** Dwarf Mouse-ear
Sepals, petals, styles, stamens in 4's or 5's	upper lvs. green		*Cerastium diffusum* Sea Mouse-ear

PORTULACEAE Purslane Family

A small easily distinguished family whose (British) flowers have 2 sepals and 5 petals.
Stamens may be 3 or 5, and stigmas 3.
The leaves are opposite and untoothed.

Radical lvs. long-stalked; only 1 pair of stem lvs.	stem lvs. fused to form a cup beneath inflor.	fls. white, under 10 mm across	*Montia perfoliata* Springbeauty
	stem lvs. distinct, unstalked	fls. pink or white, 15–20 mm across; petals notched	*Montia sibirica* Pink Purslane
Lvs. in pairs, spoon-shaped	fls. white, about 3 mm across; stamens 3	low, straggling plant of wet places	*Montia fontana* Blinks

AIZOACEAE Mesembryanthemum Family

A succulent plant naturalized on cliffs in S.W. England. The flowers are many-petalled and gaudy.

Lvs. fleshy, triangular in cross-section, 7–10 cm long	fls. purple or yellow, about 5 cm across	*Carpobrotus edulis* Hottentot-fig

AMARANTHACEAE Amaranth Family

Plants with greenish flowers in dense spikes, mixed with bristles. The leaves are stalked and ovate.

Plant grey-green, downy, 20–80 cm tall	a casual	*Amaranthus retroflexus** Common Amaranth

CHENOPODIACEAE Goosefoot Family

There are many rather dull-looking plants in this family. They all have greenish flowers. The leaves are usually alternate and sometimes rather mealy (flowry).

NOTE Plants of the Genus Chenopodium are particularly difficult to determine. This key is worked out largely on leaf shape, and this may vary considerably. A more certain identification needs ripe seeds and a microscope.

Succulent plants of tidal mud, w. no apparent lvs. (The species of this genus are not well defined)	stems w. woody base and underground runners	in firm mud	*Salicornia perennis* Perennial Glasswort
	flowers mostly in ones cf. [a]	in firm mud	*Salicornia pusilla* Fragile Glasswort
	terminal spike usu. over 10 cm w. 15–40 joints	in soft mud	*Salicornia dolichostachya* Bushy Glasswort
	terminal spike under 2 cm w. 4–6 joints	in firm mud; often tinged red	*Salicornia ramosissima* Twiggy Glasswort
	terminal spike about 2 cm w. 8–16 joints		*Salicornia europaea* Glasswort

fls. in 3's

a

93

Almost woody shrubs	lvs. linear, fleshy, 5–18 mm long		on shingle	*Suaeda vera** Shrubby Sea-blite
	lvs. elliptical, greyish		saltmarsh edges	*Halimione portulacoides* Sea-purslane
Lvs. linear, fleshy; coastal plants	lvs. 2–4 cm, spine-tipped		coastal	*Salsola kali* Saltwort
	lvs. up to 25 mm		coastal	*Suaeda maritima* Annual Sea-blite
Fr. enclosed by 2 bracts [a]	lvs. narrow w. few of small teeth [b] or [c]		lower lvs. w. 2 larger teeth [c]	*Atriplex patula* Common Orache
			on shingle and saltmarshes [b]	*Atriplex littoralis* Grass-leaved Orache
	lvs. v. frosted or silvery [d]		seashores	*Atriplex laciniata** Frosted Orache
		inflor. lfy. nearly to end	lower lvs. triangular; maritime only [e]	*Atriplex glabriuscula* Babington's Orache
			some lvs. w. 2 lobes pointing forward [c]	*Atriplex patula* Common Orache
		lower lvs. w. cuneate base	lf. lobes point forward [c]	*Atriplex patula* Common Orache
			lf. lobes point outward [f]	*Atriplex prostrata* Spear-leaved Orache
		lower lvs. w. lobes pointing outward [f]		*Atriplex prostrata* Spear-leaved Orache

Fr. w. thick perianth, often in clusters	lvs. shiny; fls. in long, narrow spikes	maritime	*Beta vulgaris* Sea Beet
Perennial w. large triangular lvs. [a]	stem base woody		[a] *Chenopodium bonus-henricus* Good King Henry
Lvs. quite or almost entire	plant mealy w. nauseous fishy smell [b]		[b] *Chenopodium vulvaria* Stinking Goosefoot
	plant hairless, odourless; lvs. ovate [c]	ripe seeds black, easily seen [c]	*Chenopodium polyspermum* Many-seeded Goosefoot
	lvs. triangular w. broad lobes [d]; sepals often 4	plant turns reddish; S.E. coasts only	[d] *Chenopodium botryodes** Small Red Goosefoot
Lower lvs. triangular w. large simple lobes [d]	some fls. w. 4 sepals	plant turns reddish; S.E. coasts only	[d] *Chenopodium botryodes** Small Red Goosefoot
Larger lvs. triangular-cordate w. a few large teeth [e]	ripe seeds black and pitted	S.E. of line Humber to Severn	[e] *Chenopodium hybridum* Maple-leaved Goosefoot
Fls. (except end ones) w. 2–4 lobed perianth	lvs. mealy beneath, bluntly toothed [f]		[f] *Chenopodium glaucum** Oak-leaved Goosefoot
	lvs. hairless both sides, well toothed [g]	plant usu. reddish	[g] *Chenopodium rubrum* Red Goosefoot
Larger lvs. 3-lobed and toothed	fl. branches almost lfless		[h] *Chenopodium ficifolium* Fig-leaved Goosefoot

Inflor. branched w. well spaced clusters		lf. [i]	*Chenopodium murale* Nettle-leaved Goosefoot
Lvs. nearly as wide as long	inflor. in dense clusters	lf. [j]	*Chenopodium opulifolium** Grey Goosefoot
30–100 cm tall; lvs. lanceolate to diamond-shaped, usu. rather mealy	v. variable, and by far the commonest Goosefoot	lf. [k]	*Chenopodium album* Fat-hen

i   j   k

Note: Do not place too much reliance on these leaf outlines. They are only an indication. *Chenopodium* leaves vary considerably.

For positive identification collect ripe seeds for microscopic examination. (See Clapham, Tutin & Warburg *Flora of the British Isles*.)

TILIACEAE   Limes

Large trees with toothed, heart-shaped leaves.
The scented yellowish flowers, which have 5 sepals, 5 petals and numerous stamens, spring in clusters from an oblong bract.

Lime fruit

Lvs. downy below	fls. usu. 3 to a cluster	lvs. 6–12 cm	*Tilia platyphyllos* Large-leaved Lime
Lvs. w. hair tufts only in vein axils and along veins	hair tufts whitish; lvs. 6–10 cm	fl. clusters (of 4–10 fls.) drooping	*Tilia × vulgaris* Lime
	hair tufts reddish; lvs. 3–6 cm	fl. clusters (of 5–10 fls.) erect or spreading	*Tilia cordata* Small-leaved Lime

MALVACEAE Mallows

Mallow flowers have 5 lightly notched and veined petals, and many stamens springing from a long tube.
Flower colour varies from palest pink to purple. Outside the 5 true sepals is an outer calyx of 3 or more segments.
Leaves are palmately lobed  or narrowly and deeply dissected.

a

Outer calyx w. 6–9 lobes	usu. not far crom the coast	*Althaea officinalis* Marsh-mallow
Outer calyx w. 3 broad lobes	tree-like, 1–3 m high	*Lavatera arborea* Tree-mallow
Upper lvs. deeply & narrowly lobed [a]		*Malva moschata* Musk Mallow
Fls. 2.5–4 cm across	petals rose-purple, striped	*Malva sylvestris* Common Mallow
Fls. up to 2.5 cm across	petals often paler, bearded at base	*Malva neglecta* Dwarf Mallow

LINACEAE Flax Family

Plants with simple, entire leaves.
The blue or white flowers have sepals, petals and stamens in 4's or 5's.

b   c

Fls. about 2 mm across; sepals 3-toothed [c]	petals 4, white; lvs. opposite	up to 8 cm tall in damp, bare places	*Radiola linoides* Allseed
Petals white; lvs. opposite			*Linum catharticum* Fairy Flax

All sepals pointed; lvs. linear, alternate, mostly w. 3 veins	fls. pale blue [b] on previous page	mostly in S.	*Linum bienne* Pale Flax
	fls. bright blue		*Linum usitatissimum* Flax
Inner sepals blunt; lvs. 1-veined, numerous	mostly in E. on calcareous grassland	fls. blue	*Linum perenne* Perennial Flax

GERANIACEAE Crane's-bills and Stork's-bills

The flowers of this family are normally some shade of pink, red, blue, or purple.
There are 5 sepals, 5 petals and 10 stamens of which only 5 may have anthers.
The leaves are alternate, with stipules, and variously lobed.

5 stamens w. anthers, and 5 without [a] (examine unopened fls.)	most lvs. twice pinnate, or nearly so.		*Erodium cicutarium* Common Stork's-bill
	most lvs. once pinnate		*Erodium moschatum* Musk Stork's-bill
	lvs. palmate		*Geranium pusillum* Small-flowered Crane's-bill
	lvs. simple w. small marginal lobes [b]	mainly S. & W. coasts	*Erodium maritimum* Sea Stork's-bill
Petals 12 mm or more long	fls. in ones, 25–30 mm across; usu. red/purple		*Geranium sanguineum* Bloody Crane's-bill
	petals black/purple w. crinkled edge		*Geranium phaeum* Dusky Crane's-bill
	petals notched	petals pale lilac w. purple veins	*Geranium versicolor* Pencilled Crane's-bill
		petals usu. purple	*Geranium pyrenaicum* Hedgerow Crane's-bill

	fr. stalks bent down	widespread	*Geranium pratense* Meadow Crane's-bill
	fr. stalks erect	not in the South	*Geranium sylvaticum* Wood Crane's-bill
Lvs. often triangular in outline, almost twice pinnate, aromatic; stem often reddish	petals 9–12 mm; pollen orange	v. common	*Geranium robertianum* Herb-Robert
	petals 6–9 mm; pollen yellow	only in South; very local	*Geranium purpureum** Little-Robin
Lvs. deeply divided into linear lobes.; lobes pinnate	fl. stalks 2–6 cm		*Geranium columbinum* Long-stalked Crane's-bill
	fl. stalks under 2 cm		*Geranium dissectum* Cut-leaved Crane's-bill
Petals distinctly notched	petals twice calyx length		*Geranium pyrenaicum* Hedgerow Crane's-bill
	petals up to 6 mm long	stem hairs long, spreading; fr. wrinkled, hairless	*Geranium molle* Dove's-foot Crane's-bill
		stem hairs v. short and close; fr. smooth, hairy	*Geranium pusillum* Small-flowered Crane's-bill
Petal margins entire	plant almost hairless; lvs. bright green	sepals erect	*Geranium lucidum* Shining Crane's-bill
	plant slightly hairy	sepals spreading	*Geranium rotundifolium* Round-leaved Crane's-bill

OXALIDACEAE Wood-sorrel Family

Plants with long-stalked trifoliolate leaves (like a large Clover leaf).
Sepals, petals and styles are 5 each and stamens 10.

Only the Wood-sorrel is a native plant.

Fls. white, in ones		*Oxalis acetosella* Wood-sorrel
Fls. yellow	stem w. long hairs; stipules small	fr. stalks bent back	*Oxalis corniculata* Sleeping Beauty
	often hairless; stipules minute or absent	fr. stalks erect or spreading	*Oxalis europaea** Upright Oxalis
Fls. pink; garden escape	lvs. all radical		*Oxalis articulata* Pink Oxalis
	some stem lvs. present; petals pale		*Oxalis incarnata** Pale Oxalis

BALSAMINACEAE Balsams

Balsam flowers have a broad lip, a hood and a spur. There are 5 stamens.
The leaves are simple, toothed, hairless and may be in 1's, 2's or 3's.
Only the Touch-me-not Balsam is a native plant.

Fls. pink/purple	lvs. lanceolate in 2's or 3's	often along river banks	*Impatiens glandulifera* Indian Balsam
Lvs. ovate w. up to 15 teeth on each side	up to 10 teeth on each side of lf.	fls. orange, blotched red; spur sharply bent	*Impatiens capensis** Orange Balsam
	10–15 teeth on each side of lf.	fls. yellow, sometimes spotted red	*Impatiens noli-tangere** Touch-me-not Balsam
Lvs. ovate w. 20 or more teeth on each side	fls. 5–15 mm, usually pale yellow; spur straight		*Impatiens parviflora* Small Balsam

ACERACEAE Maple Family

Trees or shrubs with opposite, palmately-lobed leaves.
The clustered flowers are yellowish-green and have 5 sepals, 5 petals and 8 stamens.
The winged fruit is propeller-like

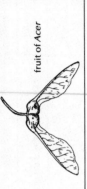

fruit of Acer

Lvs. downy below, 4–7 cm across [a]	twigs downy	milky juice in lf. stalk	*Acer campestre* Field Maple
Lvs. almost glabrous, 5–16 cm across	lvs. often shining; lf. lobes w. a few large teeth [b]	milky juice in lf. stalk	*Acer platanoides* Norway Maple
	lf. lobes irregularly toothed [c]	no milky juice	*Acer pseudoplatanus* Sycamore

c

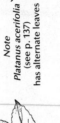

Note
Platanus acerifolia
(see p. 137)
has alternate leaves

a

b

HIPPOCASTANACEAE Horse-chestnut Family

A large tree with opposite palmate leaves.
Each inflorescence is a cluster of creamy white flowers.

Lvs. w. 5–7 distinct lflets.	fls. white, yellow-spotted	fr. a 'conker' [Sweet Chestnut is on p. 000]	*Aesculus hippocastanum* Horse-Chestnut

AQUIFOLIACEAE Holly Family

An evergreen tree or shrub.
Deciduous (male & female flowers on different trees).
The flowers have 4 petals and 4 stamens (male) or none (female).
Only the female plants will bear berries.

Most lvs. spiny	fls. white w. 4 petals	fr. a red berry

Ilex aquifolium
Holly

CELASTRACEAE Spindle Family

A deciduous shrub bearing greenish flowers and brightly coloured fruit.

Lvs. opposite, w. fine teeth; twigs green	petals and stamens 4	fr. 4-lobed, orange and pink

Euonymus europaeus
Spindle

BUXACEAE Box Family

Male and female flowers are separate, but on the same plant.
Evergreen shrub or tree, only native on chalk or limestone.

Fls. greenish white, without petals	stamens 4 or none	lvs. opposite, shiny, up to 25 mm

Buxus sempervirens
Box

RHAMNACEAE Buckthorns

Deciduous shrubs or small trees with greenish flowers and fleshy (berry-like) fruit.

Lvs. finely toothed; petals 4; stamens 4 or none [a]	fr. turns from green to black	thorny shrub on calcareous soils

Rhamnus catharticus
Buckthorn

Lf. margins entire; fls. w. 5 petals [b]	fr. turns from green to red to black	not thorny; on acid soils

Frangula alnus
Alder Buckthorn

LEGUMINOSAE (PAPILIONACEAE)   Pea Family

A large family, all (in Britain) having typical Pea-type flowers [a], which may be quite obvious, or [as in many Clovers) small and packed into tight heads [b].
Each flower has 5 sepals, 5 petals and 10 stamens.
The leaves are usually pinnate [c] or trifoliolate [d] and may bear leaf-like stipules [e].

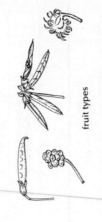

flower
standard
wings
keel
a

clover
inflorescence
b

leaf and leaflet arrangements
c   d   e
stipule

fruit types

Shrubby perennials w. woody stems	Section 1 page 103
Flowers yellow to dark cream (but not white, or creamy/dirty white)	Section 2 page 104
Leaves pinnate, with 2 or more pairs of leaflets (not counting the stipules at the base)	Section 3 page 106
Other plants, including most of the clovers	Section 4 page 108

LEGUMINOSAE   Section 1   Shrubby plants

Fls. pink; a low shrub	usu. spiny; stem hairs in 2 rows	*Ononis spinosa* Spiny Restharrow
	not usu. spiny; stem hairy all round	*Ononis repens* Common Restharrow

103

▶

Plant spiny	spines and lvs. differ from each other		*Genista anglica* Petty Whin
	bract at fl. base wider than stem [a]	may flower from Nov. to July	*Ulex europaeus* Gorse
a. bract at least 2 mm wide b. bract under 1 mm wide	spines stiff; standard over 12 mm; bract [b]	mostly in W. half of Br. Is.; fls. July to Nov.	*Ulex gallii* Western Gorse
	spines weak; standard under 12 mm; bract [b]	mostly in S.E.; fls. July to Nov.	*Ulex minor* Dwarf Gorse
Fls. blue/purple	by Scottish rivers		*Lupinus nootkatensis** Nootka Lupin
Lvs. pinnate; 3–6 lflets. each side	fls. yellow		*Colutea arborescens** Bladder-senna
Lvs. palmate	fls. yellow or white	near coast mainly in S.	*Lupinus arboreus** Tree Lupin
Some lvs. trifoliate; usu. 1–2 m high			*Cytisus scoparius* Broom
All lvs. simple; plant up to 70 cm			*Genista tinctoria* Dyer's Greenweed

LEGUMINOSAE Section 2 Herbs with yellow or creamy yellow flowers

Most lvs. pinnate w. 3 or more pairs of lflets.	lf. stalk longer than inflor.	fls. greenish–cream	*Astragalus glycyphyllos* Wild Liquorice
	fls. in ones, pale yellow	lvs. w. tendrils	*Vicia lutea* Yellow-vetch
	fls. in dense heads; sepals woolly		*Anthyllis vulneraria* Kidney Vetch
	5–8 fls. in a loose head		*Hippocrepis comosa* Horseshoe Vetch

Plant w. tendrils	stipules v. broad; [a] lflets. absent	uncommon	*Lathyrus aphaca* Yellow Vetchling
	lflets. lanceolate; some tendrils forked [b]		*Lathyrus pratensis* Meadow Vetchling
Fls. in long spikes	fls. 2–3 mm; standard longer than wings		*Melilotus indica* Small Melilot
	fls. about 5 mm; keel sl. shorter than wings [c]	ripe pods brown, glabrous	*Melilotus officinalis* Ribbed Melilot
	fls. about 5 mm; all petals equal	ripe pods black, downy	*Melilotus altissima* Tall Melilot
l.flet. margins quite entire	fl. head compact, ovoid w. many fls.	fls. pale yellow	*Trifolium ochroleuchon* Sulphur Clover
	fls. 25 mm, in ones	in S.	*Tetragonolobus maritimus** Dragon's-teeth
	calyx teeth spreading in bud	plant usu. hairy; 4–10 fls. in a head	*Lotus uliginosus* Greater Bird's-foot-trefoil
	lflets. narrowly lanceolate, usu. hairless	fls. about 10 mm, 1–4 in a head	*Lotus tenuis* Narrow-leaved Bird's-foot-trefoil
	plant hairy; fls. about 8 mm, 1–4 in head	pod up to 12 mm; S.W. coasts	*Lotus subbiflorus** Hairy Bird's-foot-trefoil
		pod over 18 mm; 1–2 fls. in a head; S. only	*Lotus angustissimus** Slender Bird's-foot-trefoil
	fls. about 15 mm	usu. hairless; common	*Lotus corniculatus* Common Bird's-foot-trefoil
Midrib of l.flet. ends w. tiny point [b] see next page	stipules toothed [b] see next page	over 8 fls. in a head; pod not spiny	*Medicago lupulina* Black Medick

stipule

a

b

standard   flower form
c

standard
wing
keel
wing bent back

fl. head about 5 mm; pods spiny [a]	lvs. often blotched; stipules coarsely toothed	*Medicago arabica* Spotted Medick
	lvs. not blotched; stipule teeth v. fine	*Medicago polymorpha* Toothed Medick
fl. head up to 25 mm; pods not spiny	plant downy	*Medicago minima* Bur Medick
		Medicago falcata Sickle Medick
Fl. head about 20 mm; pale yellow	fls. about 15 mm	*Trifolium ochroleucon* Sulphur Clover
Fl. head creamy yellow, 10–20 mm, on lfless stem	fls. 8–10 mm	*Trifolium repens* White Clover
About 40 fls. in a head		*Trifolium campestre* Hop Trefoil
2–6 fls. in a head	standard notched at tip [c]	*Trifolium micranthum* Slender Trefoil
Usu. 8–20 fls. in a head	standard not notched [d]	*Trifolium dubium* Lesser Trefoil

LEGUMINOSAE Section 3 Herbs with pinnate leaves and white/pink/blue/purple flowers

Lf. stalk ends w. a terminal lflet, not a tendril or point	fls. 3–4 mm, up to 6 in a cluster	fls. white w. red veins	*Ornithopus perpusillus* Bird's-foot
	fls. pink w. darker veins, in a spike	*Onobrychis viciifolia* Sainfoin
	stem & lvs. downy	fls. violet	*Astragalus danicus* Purple Milk-vetch
	fls. cream/green	lvs. longer than inflor.; lflets large	*Astragalus glycyphyllos* Wild Liquorice

10–20 fls. in a spherical head		fls. pink/purple/white	*Coronilla varia* Crown Vetch
inflor. oblong		fls. white/lilac	*Galega officinalis* Goat's-rue
Stem winged or clearly angled	lf. stalks in end in sharp point, not a tendril	fls. red to blue	*Lathyrus montanus* Bitter Vetch
	fls. 2–6 in a cluster; lvs. w. 10–16 lflets	fls. purple; v. common	*Vicia sepium* Bush Vetch
	fls. in ones or twos; lvs. w. 2–4 lflets	standard purplish blue, wings white	*Vicia bithynica** Bithynian Vetch
	fls. 5–15 in a cluster; lvs. w. 6–10 lflets	fls. purple; shingle plant	*Lathyrus japonicus** Sea Pea
	fls. 2–6 in a cluster; lvs. w. 4–6 lflets	fls. pale purple; marsh plant	*Lathyrus palustris** Marsh Pea
No tendril or lf. at end of lf. stalk	lvs. w. 12–20 lflets	fls. white, tinged purple; in W. & N.	*Vicia orobus* Wood Bitter-vetch
10–30 bluish fls. in long inflors.	usu. a hedgerow climber		*Vicia cracca* Tufted Vetch
Fls. whitish purple to pale yellow	fls. pale yellow, usu. in ones	fls. about 20 mm	*Vicia lutea* Yellow-vetch
	fls. tiny up to 6 together	fl. about 4 mm	*Vicia hirsuta* Hairy Tare
Fls. 2-coloured, white & blue/ purple	fls. 1 or 2 together; 1–3 pairs of lflets		*Vicia bithynica** Bithynian Vetch
	6–8 fls. in inflor.; 6–10 pairs of lflets		*Vicia sylvatica* Wood Vetch
	5–15 fls. in inflor.; 3–4 pairs of lflets	on shingle	*Lathyrus japonicus** Sea Pea

▶

Inflor. of 1–4 blue fls. on long stalk	
1 or 2 fls. together; usu. 4 seeds in a pod	*Vicia tetrasperma* Smooth Tare
up to 4 fls. together; 5–8 seeds in a pod	*Vicia tenuissima** Slender Tare
Fls. 5–8 mm, in ones	lvs. w. 1–3 pairs lflets; pods hairless	*Vicia lathyroides* Spring Vetch
Fls. 10–30 mm, in ones or twos	the more robust plants are usu. cultivated varieties	*Vicia sativa* Common Vetch
2–6 fls. in a cluster	fls. bluish purple	*Vicia sepium* Bush Vetch

LEGUMINOSAE Section 4 Other plants

Lvs. grass-like	fls. red, in ones or twos	*Lathyrus nissolia* Grass Vetchling
Lvs. w. only 2 lflets and a tendril; fls. pink	fls. 20–30 mm; lvs. ovate	*Lathyrus latifolius* Broad-leaved Everlasting-pea
	fls. about 16 mm lflets linear/lanceolate	*Lathyrus sylvestris* Narrow-leaved Everlasting-pea
Lvs. palmate	by Scottish rivers	*Lupinus nootkatensis* Nootka Lupin
Fls. white, in spikes 2–5 cm long		*Melilotus alba* White Melilot
Plant 30–60 cm high; inflor. loose	fls. bluish purple	*Medicago sativa* Lucerne
Fls. white/pink, up to 6 in a head	pod longer than calyx; 1–3 fls. in a head	*Trifolium ornithopodioides* Fenugreek
	pods bury themselves in the ground	*Trifolium subterraneum* Subterranean Clover

Fl. heads up to 10 mm on v. short stalks	petals twice as long as calyx teeth	lvs. downy; fls. white flushed w. pink	*Trifolium striatum* Knotted Clover
		lvs. hairless; fls. pale purple	*Trifolium glomeratum* Clustered Clover
	lvs. & calyx hairy	fls. white	*Trifolium scabrum* Rough Clover
	lvs. & calyx hairless; fls. close to ground	fls. white; mainly S. & E. coasts	*Trifolium suffocatum** Suffocated Clover
Sepals v. hairy, longer than petals			*Trifolium arvense* Hare's-foot Clover
Fls. crimson			*Trifolium incarnatum* Crimson Clover
Fl. heads on lfless stalks rising from creeping runner	lvs. usu. w. white blotches; fls. usu. whitish		*Trifolium repens* White Clover
	fls. pink; pods enclosed by calyx	fr. head strawberry-like	*Trifolium fragiferum* Strawberry Clover
Fl. heads stalkless; often in lf. axils	fls. white, flushed w. pink		*Trifolium striatum* Knotted Clover
Standard much longer than calyx	calyx white w. green points	fls. pink/white; all fl. heads axillary	*Trifolium hybridum* Alsike Clover
		fls. purple; fl. heads terminal	*Trifolium medium* Zigzag Clover
	one pair of lvs. v. close to fl. head	calyx usu. downy	*Trifolium pratense* Red Clover
	calyx hairless; inflor. stalk w'out lvs.	fls. reddish purple	*Trifolium medium* Zigzag Clover
Fls. pink, up to 7 mm	in saltmarshes		*Trifolium squamosum* Sea Clover

standard

calyx

ROSACEAE Rose Family

A large family containing plants of very varied appearance.
The flowers normally have 5 sepals and petals and more than 10 stamens.
Leaves are usually alternate, with stipules.

Other families which have many-stamened flowers are:
RANUNCULACEAE p. 68 (Buttercup family)
GUTTIFERAE p. 85 (St John's-worts) – but these have opposite leaves.

Trees, shrubs, or woody scramblers		Section 1 page 110
Flowers with yellow petals		Section 2 page 112
Other plants		Section 3 page 114

ROSACEAE Section 1 Trees, shrubs and woody scramblers

Fls. appear before lvs. in early Spring	a thorny shrub	fls. white; fr. black	*Prunus spinosa* Blackthorn
Evergreen shrub w. small, dark lvs. (up to 3 cm)	upright plant; lvs. 2–3 cm	garden escape	*Cotoneaster simonsii* Himalayan Cotoneaster
	prostrate plant; lvs. up to 1 cm	garden escape	*Cotoneaster microphyllus* Small-leaved Cotoneaster
Scrambling, prickly plants (Roses & Brambles)	fr. soft & juicy	stem w. many tiny prickles, fr. red	*Rubus idaeus* Raspberry
		fr. w. up to 5 segments, and whitish 'bloom'; sepals upturned	*Rubus caesius* Dewberry
		very variable	*Rubus fruticosus* (agg.) Bramble
	stem densely prickly; all sepals entire	fls. cream; fr. black	*Rosa pimpinellifolia* Burnet Rose

Stigmas of Roses	stigmas on top of one long style [a]	fls. always white; sepals w. few or small lobes [d]	*Rosa arvensis* Field Rose
		lvs. v. glandular beneath / prickles hooked	*Rosa rubiginosa* (agg.) Sweet Briar
	stigmas united on a short style [b]	some sepals w. long lobes [e]	*Rosa stylosa* Wild Rose
	stigmas free [c]	lvs. downy both sides; petals deep pink	*Rosa tomentosa* (agg.) Downy Rose
		lvs. hairless above; prickles well curved	*Rosa canina* (agg.) Dog Rose
Stems upright, usu. unbranched	prickles tiny & many		*Rubus idaeus* Raspberry
Bush w. small pink fls. in dense heads	lvs. w'out stipules		*Spirea salicifolia* Bridewort
Tree w. pinnate lvs.	fls. in large clusters		*Sorbus aucuparia* Rowan
Lvs. lobed or doubly toothed	thorny	fls. w. 1 style	*Crataegus monogyna* Hawthorn
		2 or 3 styles	*Crataegus laevigata* Midland Hawthorn
	lvs. v. white beneath; usu. over 10 pairs of veins	fr. red; lf. as [f]	*Sorbus aria* Common Whitebeam
	lvs. grey beneath	fr. red; lf. as [g]	*Sorbus intermedia* Swedish Whitebeam
	older lvs. green both sides [i]	fr. orange/brown; lf. as [h]	*Sorbus latifolia** Broad-leaved Whitebeam
		lf. as [i]	*Sorbus torminalis* Wild Service-tree

10-30 fls. together in loose spikes; ripe fr. purple/black	petals about 5 mm long		*Prunus padus* Bird Cherry
	petals 15 mm or more	young lvs. copper-coloured	*Amelanchier lamarckii** Juneberry
fls. w. 5 styles	anthers purple		*Pyrus pyraster* Wild Pear
	anthers yellow	calyx & older lvs. hairless beneath	*Malus sylvestris* Crab Apple
		calyx & older lvs. downy beneath	Cultivated Apple
Fls. 2-6 in a cluster w. lf.-like scales at base of inflor.	lvs. pale green, dull, downy beneath	usu. a tree; lf. stalks have 2 red knobs (glands)	*Prunus avium* Wild Cherry
	lvs. dark green, shiny, often hairless beneath	usu. a shrub	*Prunus cerasus* Dwarf Cherry
Fls. solitary, or up to 3 in a cluster	lvs. & twigs usu. dull, often downy		*Prunus domestica* Wild Plum
	lvs. & twigs often glossy	fr. yellow or red	*Prunus cerasifera* Cherry Plum

ROSACEAE Section 2 Herbs with yellow flowers

Inflor. a tall tapering spike	fr. w. long grooves	outer hooks on fr. spread forward [a]	*Agrimonia eupatoria* Agrimony
	fr. hardly grooved; lvs. w. many glands beneath	outer hooks on fr. spread backward [b]	*Agrimonia procera* Fragrant Agrimony
Radical lvs. pinnate w. larger end lflet; fr. hooked	fls. erect; calyx green		*Geum urbanum* Wood Avens
	fls. nodding; calyx purple		*Geum rivale* Water Avens

Lvs. pinnate, often silvery beneath			*Potentilla anserina* Silverweed
Lvs. white beneath, dark green above			*Potentilla argentea* Hoary Cinquefoil
Fls. mostly w. 4 petals	lvs. mostly stalked w. 3 lflets		*Potentilla anglica* Trailing Tormentil
	lvs. mostly stalkless w. 5 lflets		*Potentilla erecta* Tormentil
Erect stiff plant; fls. 20–25 mm	petals notched		*Potentilla recta** Sulphur Cinquefoil
Creeping plant w. solitary fls. and stalked lvs.	petals 5, v. narrow; lflets in 3's usu. 3-toothed	small compact plant of highlands in N.	*Sibbaldia procumbens* Sibbaldia
	most lvs. w. 5 lflets; most fls. w. 5 petals		*Potentilla reptans* Creeping Cinquefoil
	lflets 3, 4 or 5; petals 4 or 5		*Potentilla anglica* Trailing Tormentil
Petals longer than sepals	fls. 15–25 mm, often orange spotted	alpine in calcareous grassland	*Potentilla crantzii** Alpine Cinquefoil
	fls. 10–15 mm across	in calcareous grassland	*Potentilla tabernaemontani* Spring Cinquefoil
Fls. in compact clusters; petals lanceolate	compact plant about 2 cm high	highlands in N.	*Sibbaldia procumbens* Sibbaldia
Lvs. w. 3 lflets and 2 stipules			*Potentilla norvegica** Ternate-leaved Cinquefoil
Lower lvs. w. 5 lflets & 2 stipules			*Potentilla intermedia* Intermediate Cinquefoil

ROSACEAE Section 3 Other plants

Fls. packed tightly into a long-stalked knob; lvs. pinnate	low creeping plant; lvs. w. 3–4 prs. lflets	fr. spiny	*Acaena novae-zelandiae** Pirri-pirri-bur
	4 stamens to each fl.; sepals maroon		*Sanguisorba officinalis* Great Burnet
	stamens numerous; sepals green	robust specimens may be cultivated escapes	*Sanguisorba minor* Salad Burnet
Fls. green in clusters, or sometimes single	sepals 5; lflets in 3's, usu. 3-toothed [a]	small plant of highlands in N.	*Sibbaldia procumbens* Sibbaldia
	fl. clusters small, stalkless, among lvs. [b]		*Aphanes arvensis* Parsley-piert
	lvs. w. long lobes, silvery beneath [c]		*Alchemilla alpina* Alpine Laidy's-mantle
	lvs. broad, green both sides		*Alchemilla vulgaris* (agg.) Lady's-mantle
Petals purple			*Potentilla palustris* Marsh Cinquefoil
Petals orange/pink			*Geum rivale* Water Avens
Lvs. pinnate	plant 60–120 cm high; fls. usu. 5-petalled		*Filipendula ulmaria* Meadowsweet
	plant up to 80 cm; 6 petals per fl.		*Filipendula vulgaris* Dropwort
Lvs. not divided into lflets	petals 8; lf. shape [d]	mt. plant	*Dryas octopetala** Mountain Avens

a b c d

a	petals 5; lf. shape [a]	mountain moors; fr. red to orange
		Rubus chamaemorus Cloudberry
Petals lanceolate	lvs. w. 3 lflets	fr. red
		Rubus saxatilis Stone Bramble
Fls. 20–35 mm		
		Fragaria ananassa Garden Strawberry
Petals lightly notched, well spaced		end tooth of lflet much shorter than its neighbours
		Potentilla sterilis Barren Strawberry
Petals rounded		end tooth of lflet not shorter than its neighbours
		Fragaria vesca Wild Strawberry

CRASSULACEAE Stonecrop Family

The leaves are simple, more or less succulent, and have no stipules. Flowers with 3, 4 or 5 sepals and petals and 0, 3, 8 or 10 stamens. All except *Umbilicus* have free petals.

Fl. parts in 3's	tiny plant of bare ground	*Crassula tillaea** Mossy Stonecrop
Corolla tubular	lvs. round, w. rounded teeth; fls. cream	*Umbilicus rupestris* Navelwort

▶

Petals 4; stamens 0 or 8	mountain plant			*Sedum rosea* Roseroot
Lvs. flat, toothed; fls. pink/red	lvs. alternate			*Sedum telephium* Orpine
	lvs. opposite	garden escape		*Sedum spurium** Caucasian Stonecrop
Petals pink; lvs. glandular	in North only			*Sedum villosum* Hairy Stonecrop
Petals white	lvs. mostly opposite; almost downy			*Sedum dasyphyllum* Thick-leaved Stonecrop
		lvs. 3–5 mm; fls. about 12 mm	inflor. usu. w. 2 main branches	*Sedum anglicum* English Stonecrop
		lvs. 6–12 mm; fls. 6–9 mm	inflor. w. several main branches	*Sedum album* White Stonecrop
Lvs. 3–5 mm, blunt; fls. yellow				*Sedum acre* Biting Stonecrop
Lower lvs. on fl. stem upright; fls. yellow	lvs. w. flat upper surface	dead lvs. persist on sterile shoots		*Sedum forsterianum* Rock Stonecrop
Lower lvs. on fl. stem spreading; fls. yellow	lvs. w. rounded upper surface	dead lvs. fall from sterile shoots		*Sedum reflexum* Reflexed Stonecrop

SAXIFRAGACEAE Saxifrages

Flowers on the genus Saxifraga have 5 sepals and petals and 10 stamens, while those of Chrysosplenium have 4 (sometimes 5) sepals, no petals and 8 (sometimes 10) stamens.
The fruit is 2-lobed.
The leaves are simple, but may be lobed or toothed.

Fls. yellowish green w'out petals; sepals usu. 4; stamens 8	in wet shady places	stem lvs. rounded, opposite [b]	*Chrysosplenium oppositifolium* Opp.-leaved Golden-Saxifrage
		lvs. alternate	*Chrysosplenium alternifolium* Alternate-leaved Golden Saxifrage
Fls. purple	Northern and mt. plant		*Saxifraga oppositifolia* Purple Saxifrage
Fls. yellow	on Northern mts.		*Saxifraga aizoides* Yellow Saxifrage
Lvs. ovate in a basal rosette	calyx bent right down in fl. & fr.	mt. districts; petals white; lf. shape [b]	*Saxifraga stellaris* Starry Saxifrage
	fl. stems downy; inflor. usu. compact	on Northern mts.; petals white; lf. shape [a]	*Saxifraga nivalis** Alpine Saxifrage
Many creeping shoots w. linear lvs.	hill districts; petals white	stem lvs.	*Saxifraga hypnoides* Mossy Saxifrage
Petals about 3 mm, white; lvs. 3-lobed	on dry, sandy ground and wall tops	lvs.	*Saxifraga tridactylites* Rue-leaved Saxifrage
Petals over 10 mm; white; lower lvs. rounded	lf shape [a]		*Saxifraga granulata* Meadow Saxifrage

117

PARNASSIACEAE Grass of Parnassus

Only one species. The petals are white with green veins. There are 5 sepals, petals and fertile stamens, but many barren stamen-like glands within the flower.
The radical leaves are cordate and long-stalked.

Only one lf. (not stalked) on each fl. stem	in wet places	*Parnassia palustris* Grass-of-Parnassus

GROSSULARIACEAE Currant Family

Shrubs with alternate, palmately lobed leaves.

The clustered greenish flowers have 5 sepals, petals and stamens, and 2 stigmas.
The fruit is an edible berry.

Plant spiny	fr. green	*Ribes uva-crispa* Gooseberry
Fls. w. stamens or stigmas but not both	fr. red	on limestone	*Ribes alpinum* Mountain Currant
Lvs. aromatic	fr. black	*Ribes nigrum* Black Currant
Lvs. not aromatic	fr. red	*Ribes rubrum* (agg.) Red Currant

DROSERACEAE Sundews

Insectivorous plants of wet heath and bog.
The small white flowers (about 5 mm across), with 5–8 petals and sepals are produced in a spike rising from a rosette of leaves which are covered with red glands (long hairs with a sticky globule on the end).

Lvs. round, in a flat rosette		*Drosera rotundifolia* Round-leaved Sundew
Lf. blades narrow, tapering, up to 3 cm	lvs. usu. erect	*Drosera anglica* Great Sundew
Lf. blade oval, about 1 cm		*Drosera intermedia* Oblong-leaved Sundew

LYTHRACEAE Purple-loosestrife Family

Two very different plants are included in the family.
The flowers usually have 6 (or no) petals and 6 or 12 stamens.
The stems are often 4-angled.

a

60–120 cm tall; fls. showy, purple, in a spike	lvs. lanceolate, in 2's or 3's	*Lythrum salicaria* Purple-loosestrife
	often at waterside	
Small creeping plant; fls. about 1 mm in lf. axils [a]	lvs. about 1 cm	*Lythrum portula* Water-purslane
	in damp, bare places	

THYMELAEACEAE Spurge-laurel

An evergreen shrub up to about 1 m tall with lanceolate leaves.
The flowers have 4 sepals, no petals and 8 stamens.

| Lvs. dark, shiny, up to 12 cm; fls. greenish | fr. black, ovoid, berry-like | favours woods on calcareous soils | *Daphne laureola* Spurge-laurel |

ELAEAGNACEAE Sea-buckthorn

A thorny shrub, up to 3 m.
Minute greenish flowers appear before the leaves, having 2 sepals but no petals.

| Lvs. linear/lanceolate, silvery up to 8 cm | fr. an orange berry | usu. maritime, may be planted | *Hippophaë rhamnoides* Sea-buckthorn |

ONAGRACEAE Willowherb Family

Circaea has 2 sepals, petals and stamens. Other members of the family have 4 sepals and petals and 8 stamens.
The leaves are lanceolate to ovate, often toothed.

The genus *Epilobium* hybridizes freely and many plants with intermediate characters can be found. *Epilobium* flowers range from very pale to dark rose, have long, narrow fruits and plumed seeds.

Shrub w. hanging red & purple fls.	sepals red, about 12 mm; stamens project	introduced hedging plant in W.	*Fuchsia magellanica* Fuchsia
Fls. yellow	stem hairs w. red bases	petals over 30 mm	*Oenothera erythrosepala* Large-flowered Evening-primrose
		petals under 30 mm	*Oenothera cambrica* Small-flowered Evening-primrose
	older fls. turn red	petals over 30 mm	*Oenothera stricta** Fragrant Evening-primrose
	fls. remain yellow	petals 20–30 mm	*Oenothera biennis* Common Evening-primrose

Sepals, petals, stamens 2 each per fl.; fr. bristly	tiny bract at base of each fl. stalk	N. & W.	*Circaea intermedia* Upland Enchanter's-nightshade
	no bract at base of fl. stalk	common	*Circaea lutetiana* Enchanter's-nightshade
Tall plant (usu. 50–150 mm) w. fls. over 15 mm across	lvs. opposite, hairy		*Epilobium hirsutum* Great Willowherb
	lvs. alternate, almost hairless		*Chamerion angustifolium* Rosebay Willowherb
Small creeping plant w. broadly ovate lvs. in pairs	fl. stalks erect		*Epilobium brunnescens* New Zealand Willowherb
Stigma 4-lobed	plant greyish, woolly below	lf. shape [e] on next page	*Epilobium parviflorum* Hoary Willowherb
	lvs. mostly opposite w. toothed & rounded base	lf. shape [c] on next page	*Epilobium montanum* Broad-leaved Willowherb
	lvs. often alternate w. untoothed cuneate base	in S.; lf. shape [e] on next page	*Epilobium lanceolatum* Spear-leaved Willowherb
Small alpine (up to 20 cm) by mt. streams in N.	stem w. 2 faint ridges	usu. under 10 cm; lvs. 1–2 cm, pale green [a]	*Epilobium anagallidifolium* Alpine Willowherb
		taller plant; lvs. up to 4 cm, dark green [b]	*Epilobium alsinifolium* Chickweed Willowherb
	stem w' out ridges; lvs. narrowly lanceolate	lf. shape [g] on next page	*Epilobium palustre* Marsh Willowherb
	lvs. distinctly stalked w. cuneate base	fls. white/pale pink	*Epilobium roseum* Pale Willowherb
		lf. shape [e] on next page	

Stems w'out lines or ridges	lvs. usu. opposite, narrow	lf. shape [g]	*Epilobium palustre* Marsh Willowherb
Many spreading and glandular stem hairs	2 or 4 raised lines on stem	lf. shape [d]	*Epilobium ciliatum* American Willowherb
Stem w. 4 raised lines	some glandular hairs on calyx base	lvs. dull w. small teeth; lf. shape [d]	*Epiloboium obscurum* Short-fruited Willowherb
	calyx w' out glands; lvs. shiny, well toothed	lf. shapes [f] + [g]	*Epilobium tetragonum* Square-stalked Willowherb

leaf shapes

c   d   e   f   g

NOTE: teeth omitted

HALORAGACEAE Water-milfoils

Submerged aquatics with feathery leaves.
The tiny green/yellow/red flowers are arranged in spikes usually rising above the surface. The upper flowers usually have 8 stamens, the lower ones 4 stigmas.

fl. spike

whorl of leaves

Lvs. much longer than internodes, usu. 5 to a whorl	fl. bracts pinnate, as long as fls.	fls. in whorls	*Myriophyllum verticillatum* Whorled Water-milfoil
Lvs. usu. 3–4 to a whorl	upper fls. 1 or 2 to a whorl	lvs. w. up to 18 segments altogether	*Myriophyllum alternifolium* Alternate Water-milfoil
	fls. usu. 4 to a whorl	all but lowest fl. bracts entire	*Myriophyllum spicatum* Spiked Water-milfoil

HIPPURIDACEAE Mare's-tail

An aquatic plant with whorls of entire, linear leaves.
The tiny greenish flowers are close to the stem.
This plant should not be confused with the Horsetails on p. 000.

		Hippuris vulgaris Mare's-tail
Lvs. 6–12 to a whorl		

CALLITRICHACEAE Water-starworts

Plants growing in water or on wet mud.
The leaves are opposite and entire varying in shape from linear to nearly round; they may be submerged, floating or aerial, and vary in shape according to position and habitat.
The minute flowers are in the leaf axils.
NOTE: The species are difficult to determine without ripe fruit, so this should be searched for.

		winged fruit
Lvs. mostly or all linear, about 2 mm wide, up to 25 mm long	a few wider lvs. at stem tips	fr. keeled, but not winged	*Callitriche hamulata* Intermediate Water-starwort
	lvs. all linear, all submerged	fr. w. 4 wings; in N. & W.	*Callitriche hermaphroditica* Autumnal Water-starwort
Fr. lobes distinctly winged & grooved	most lvs. broad about 10 × 7 mm	v. common	*Callitriche stagnalis* Common Water-starwort
Fr. lobed but not winged	lower lvs. often linear		*Callitriche platycarpa* (agg.) Various-leaved Water-starwort

123

LORANTHACEAE Mistletoe

A semi-parasitic woody evergreen, growing most often on deciduous trees, especially apple, poplar and lime.

Lvs. oblong, in terminal pairs; fls. tiny, greenish, in small clusters	in fl. Feb.–May	fr. a white berry	*Viscum album* Mistletoe

SANTALACEAE Bastard-toadflax

A low creeping plant with alternate leaves and tiny star-like flowers in loose spikes. Sepals and stamens usually 5, petals absent.

Fls. 3–4 mm, green & white, each w. 3 bracts	lvs. linear	in calcareous turf	*Thesium humifusum** Bastard-toadflax

CORNACEAE Dogwood Family

The flowers have sepals, petals and stamens in 4's. The leaves are untoothed and opposite.

Fls. about 2 mm, purple, in tight cluster, surrounded by 4 white bracts	up to 20 cm tall	moors in Scotland and N. England	*Cornus suecica* Dwarf Cornel
Shrub, often w. dark red twigs	fls. white; berries black	in scrub, often on calcareous soil	*Cornus sanguinea* Dogwood

ARALIACEAE Ivy

An evergreen woody climber, with dark-green variably-shaped leaves.

Fls. yellow/green, late in year	fr. a black berry

Hedera helix
Ivy

UMBELLIFERAE Umbellifers

Nearly all the plants of this family bear their flowers in a characteristic umbel.
Each flower has 5 petals, 5 stamens and 2 stigmas. The calyx has either 5 small teeth or none at all.
At the base of each main umbel there may be bracts, and at the base of the secondary umbels there may be bracteoles (see diagram).
The leaves are normally alternate and may be repeatedly divided into pinnate lobes (see diagram)
The 2-lobed fruits are very often helpful for identification.
There is a Simplified Key to common white umbellifers on page 255.
The B.S.B.I. handbook *Umbellifers of the British Isles* contains comprehensive keys, illustrations and descriptions.

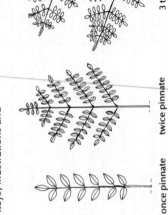

an umbel

flower
bracteole
ray
bract

once pinnate   twice pinnate   3 times pinnate

Lvs. round, attached at centre to stalk	low, creeping marsh plant; umbels minute

Hydrocotyle vulgaris
Marsh Pennywort

▶

Fls. greenish-yellow	lvs. v. succulent w. linear lobes [a]	on maritime rocks	*Crithmum maritimum* Rock Samphire
	lvs. simple; inflor. slender	near coast	*Bupleurum tenuissimum** Slender Hare's-ear
	lvs. v. finely divided [b] or [c]	aromatic; lf. segments almost hair-fine [b]	*Foeniculum vulgare* Fennel
		lf. segments flattened [c]	*Silaum silaus* Pepper-saxifrage
	plant v. aromatic	smelling of celery; lflets broad, shiny [d]	*Apium graveolens* Wild Celery
		stem hollow; lvs. usu. once pinnate [e]	*Pastinaca sativa* Wild Parsnip
		stem solid; lvs. usu. twice pinnate	*Petroselinum crispum* Garden Parsley
		lflets mostly in 3's; ripe fr. black	*Smyrnium olusatrum* Alexanders
	lf. segments linear/lanceolate [c]		*Silaum silaus* Pepper-saxifrage
lvs. w. strong spines	fls. blue	on maritime sands	*Eryngium maritimum* Sea-holly
Radical lvs. palmately lobed		woodland plant	*Sanicula europaea* Sanicle
Lower aerial lvs. once pinnate (lflets may be deeply toothed)	petal margins entire	submerged lvs. v. finely dissected	*Apium inundatum* Lesser Marshwort
N.B. Ignore submerged lvs. if plant is aquatic.		bracteoles present	*Apium nodiflorum* Fool's Water-cress
		no bracteoles; smells of celery	*Apium graveolens* Wild Celery

umbel of 3–6 rays of irregular length	smells of parsley; bracts see [a]		*Petroselinum segetum* Corn Parsley
	smell unpleasant; bracts see [b]		*Sison amomum* Stone Parsley
usu. growing in water	umbels terminal		*Sium latifolium* Great Water-parsnip
	umbels [c] opposite lf. or in lf. axil		*Berula erecta* Lesser Water-parsnip
outer petals much larger than inner	umbels 30 cm or more across; often 3 m high		*Heracleum mantegazzianum* Giant Hogweed
	50–200 cm high		*Heracleum sphondylium* Hogweed
some lvs. twice pinnate; no bracts	stem tough, downy		*Pimpinella saxifraga* Burnet-saxifrage
all lvs. once pinnate; no bracts	stem hairless, brittle		*Pimpinella major* Greater Burnet-saxifrage
Purple spots on stem	stem rough, up to 1 m		*Chaerophyllum temulentum* Rough Chervil
	stem smooth, up to 2 m		*Conium maculatum* Hemlock
Fr. bristly	bracts many, divided into long segments		*Daucus carota* Wild Carrot
	umbels almost stalkless		*Torilis nodosa* Knotted Hedge-parsley
	bracts 4–12	fr. w. hooked spines [d]	*Torilis japonica* Upright Hedge-parsley

a

b

c

d

▶

bracts 0 or 1	fr. rounded at top [a]; stem solid		*Torilis arvensis* Spreading Hedge-parsley
	fr. beaked [b]; stem hollow		*Anthriscus caucalis* Bur Chervil
Fr. slender, 3–7 cm long	rays 1–3; bracts 0 or 1; bracteoles about 5		*Scandix pecten-veneris* Shepherd's-needle
Outer bracteoles turned down in 3's			*Aethusa cynapium* Fool's Parsley
Stem partly suffused with purple	lvs. twice trifoliate, glossy	up to 90 cm high; on northern coasts	Ligusticum scoticum Scots Lovage
	lvs. twice pinnate; lflets broad	in fl. July–Sep.; up to 2 m high	*Angelica sylvestris* Wild Angelica
	lvs. almost 3 times pinnate	in fl. March–June	*Anthriscus sylvestris* Cow Parsley
Stem solid	lvs. fine, in whorls [c]	in N. & W.	*Carum verticillatum* Whorled Caraway
	No bracts or bracteoles		*Pimpinella saxifraga* Burnet-saxifrage
	rays 3–6, unequal		*Sison amomum* Stone Parsley
	styles on fr. recurved [d]	in Herts. & Cambs.	*Bunium bulbocastanum** Great Pignut
	styles on fr. erect or spreading	bracteoles subulate; in S. only, fr. as [e]	*Oenanthe pimpinelloides* Corky-fruited Water-dropwort
		bracteoles lanceolate in marshy places; fr. as [f]	*Oenanthe lachenalii* Parsley Water-dropwort

a b

c d e f

fruits

Lvs. once or twice 3-lobed; no bracts	bracteoles 0; rays 15–20		*Aegopodium podagraria* Ground-elder
	bracteoles few; rays 20–50		*Peucedanum ostruthium** Masterwort
Lvs. up to 60 cm, rough; outer petals v. large			*Heracleum sphondylium* Hogweed
Stems smooth or striate but not ridged	aquatic or marsh plants	under 30 cm high; submerged lvs. often finely divided	*Apium inundatum* Marshwort
		erect plant; umbel of 2–5 stout rays; fr. as [a]	*Oenanthe fistulosa* Tubular Water-dropwort
		submerged lvs. w. cuneate segments; fr. as [b]	*Oenanthe fluviatilis* River Water-dropwort
		submerged lvs. v. finely divided; fr. as [c]	*Oenanthe aquatica* Fine-leaved Water-dropwort
		styles on fr. erect	common	*Conopodium majus* Pignut
		lf. segments almost thread-like	mt. districts	*Meum athamanticum* Spignel
		lf. segments linear	fr. aromatic	*Carum carvi* Caraway
4 or more bracts (which may fall early)	upper lf. lobes few, v. narrow; up to 15 rays	in S. only	*Oenanthe pimpinelloides* Corky-fruited Water-dropwort
	styles long, erect [d]	common	*Oenanthe crocata* Hemlock Water-dropwort
	styles short; fr. winged [e]	in marshes	*Peucedanum palustre** Milk-parsley

▶

Larger lvs. 3 times pinnate	fr. smooth, 4–6 mm	12 or more rays; in fl. June–Sep.	*Oenanthe crocata* Hemlock Water-dropwort
		4–10 rays; in fl. March–June	*Anthriscus sylvestris* Cow Parsley
	fr. ribbed, 20–25 mm; lvs. aromatic	in North only	*Myrrhis odorata* Sweet Cicely
Lvs. once or twice pinnate; bracts few or none	lflets toothed; umbel w. 10–30 rays; fr. as [a]	 a	*Cicuta virosa** Cowbane
	fr. cylindrical [b] umbel w. 4–8 rays	b by fresh water	*Oenanthe silaifolia** Narrow-leaved Water-dropwort
	fr. ovoid 6–12 rays	in fields and woods	*Conopodium majus* Pignut

CUCURBITACEAE White Bryony

A climbing plant with tendrils and palmately lobed leaves.
The flowers have 5 sepals and petals and either 5 stamens or 3 stigmas.

Lvs. palmately lobed; fls. pale yellowish-green	ripe fr. a red berry	*Bryonia dioica* White Bryony

EUPHORBIACEAE Spurge Family

The yellowish-green flowers of the true Spurges are mostly in a terminal umbel which often has a whorl of leafy bracts at its base. Each apparent 'flower' is set in a cup which has 4 lobes (either rounded or crescent-shaped). Inside the cup are a number of stamens and one female flower whose 3-styled fruit projects on a long stalk (see [a] and [b]). Below these cups are pairs of bracts, sometimes saucer-like [c].
The leaves of Spurges are almost always alternate and simple, and the stems contain a milky juice.
The two species of Mercury have opposite, toothed leaves and small green flowers in loose spikes [d].

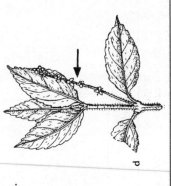

flowers of *Euphorbia*

a

b

species on p. 132 have horned lobes like this

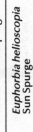

c

d

Fls. w. 3 sepals; no milky juice in the stem	hairy unbranched plant; inflor. [d]	often abundant in woodland	*Mercurialis perennia* Dog's Mercury
	branched	in waste places	*Mercurialis annua* Annual Mercury
Lvs. or lf.-like bracts opposite 	over 30 cm high; umbels 2–6 rayed	usu. a garden escape	*Euphorbia lathyrus* Caper Spurge
	under 30 cm high; umbels 1–3 rayed	true lvs. spirally arranged	*Euphorbia peplus* Petty Spurge
Lobes of flower cup rounded [a]; lvs. finely toothed 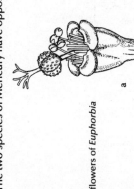	lvs. lanceolate; fr. warty 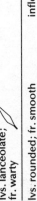		*Euphorbia platyphyllos** Broad-leaved Spurge
	lvs. rounded; fr. smooth	inflor. [c]	*Euphorbia helioscopia* Sun Spurge

▶

Lobes of fl. cup crescent-shaped [b]	bracts beneath fl. heads joined in pairs [a]		
	umbel usu. w. 8 or more rays; lvs. linear	stems downy; woodland plant	*Euphorbia amygdaloides* Wood Spurge
		stems smooth; lvs. often overlapping; in coastal sand	*Euphorbia paralias* Sea Spurge
		up to 30 cm high; lvs. about 2 mm wide	*Euphorbia cyparissias** Cypress Spurge
		30–80 cm high; lvs. 4–6 mm wide	*Euphorbia esula* Leafy Spurge
	Lvs. many; umbel 3–8 rayed	umbel usu. 3-rayed; weed of waste/cultivated places	*Euphorbia peplus* Petty Spurge
		lvs. minutely pointed w. obvious midrib; S. & W. coasts	*Euphorbia portlandica* Portland Spurge
		lvs. w' out point; midrib obscure; in coastal sand	*Euphorbia paralias* Sea Spurge
	lvs. few, linear; umbel usu. 3-rayed		*Euphorbia exigua* Dwarf Spurge

POLYGONACEAE Dock Family

A family of green, white or pink flowers, often in spikes.
The leaves are alternate and simple, many having a thin, often whitish, sheath at their base.
In *Polygonum* the perianth is petal-like, usually of 5 segments.
In *Rumex* the perianth is sepal-like, in 2 groups of 3. The 3-faced fruits of *Rumex* may be needed for identification.
There is a Simplified Key to common Docks on p. 262.
The B.S.B.I. handbook *Docks and Knotweeds of the British Isles* contains comprehensive keys, illustrations and descriptions.

leaf shapes

a b c  d e

fr.

Plant w. twining stems	stems angled, up to about 1 m long	fr. stalks up to 3 mm; fr. dull black lf. [a]	*Fallopia convolvulus* Black-bindweed
	stems round, often over 1 m long	fr. stalks 5–8 mm; fr. shiny	*Fallopia dumetorum** Copse-bindweed
Lvs. broadly & sharply triangular		lf. shape [b]	*Fagopyrum esculentum* Buckwheat
Radical lvs. long-stalked, kidney-shaped, 1–3 cm [c] 	straggling or prostrate mt. plant	lf. shape [c]	*Oxyria digyna* Mountain Sorrel
Fls. one to few together along stem	fr. distinctly longer than perianth	on shores; lf. shape [d]	*Polygonum oxyspermum** Ray's Knotgrass
	fr. not or hardly longer than perianth	v. common; lf. shape [e]	*Polygonum aviculare* Knotgrass

▶

Fls. 5-lobed, white/pink

lf. shapes

a   b   c   d   e

f   g

lf. shape

sturdy almost woody stems, over 1 m tall		up to 2 m tall; lvs. up to 12 cm, cuspidate [a]	*Reynoutria japonica* Japanese Knotweed
		2–3 m tall; lvs. cordate over 15 cm long [b]	*Reynoutria sachalinensis* Giant Knotweed
	lf. shape [c]	sometimes aquatic	*Polygonum amphibium* Amphibious Bistort
plant w. only one spike		fl. spike w. bulbs at its base; hills in N. lf. [e]	*Polygonum viviparum* Alpine Bistort
		stamens 8; styles 3 lf. [d]	*Polygonum bistorta* Common Bistort
		stamens 5; styles 2 lf. [e]	*Polygonum amphibium* Amphibious Bistort
minute yellow dots (glands) on fl. stalks and/or perianth		inflor. narrow, drooping; glands on perianth only	*Polygonum hydropiper* Water-pepper
		glands on fl. stalks	*Polygonum lapathifolium* Pale Persicaria
inflor. stout, dense		lvs. often blotched [f]	*Polygonum persicaria* Redshank
lvs. 10–25 mm wide			*Polygonum mite** Tasteless Water-pepper
lvs. usu. under 10 mm wide		lf. shape [g]	*Polygonum minus** Small Water-pepper
Slender plant up to 30 cm high			*Rumex acetosella* Sheep's Sorrel

Primary character	Secondary character	Notes / distribution	Species
Upper lvs. w. lobes clasping stem — lf. shape			*Rumex acetosa* Common Sorrel
Ripe fr. w. long teeth — fr. w. 1 wart; fr. w. 3 warts (a, b, c)	some branches almost horizontal	fl. whorls small, well spaced; lf. shape [a]	*Rumex pulcher* Fiddle Dock
	fr. usu. w. only one wart [b]	v. common	*Rumex obtusifolius* Broad-leaved Dock
	some fr. teeth longer than fr.	inland & maritime	*Rumex maritimus* Golden Dock
	teeth shorter than fr. [c]	mostly in E.	*Rumex palustris* Marsh Dock
Lvs. almost as wide as long	inflor. dense, erect	mostly in N.	*Rumex alpinus* Monk's-rhubarb
Ripe fr. w'out warts	in N. only		*Rumex longifolius* Northern Dock
Fl. whorls dense & close; fr. w'out teeth (d)	lvs. w. wavy edges; fr. [d]	v. common	*Rumex crispus* Curled Dock
	fr. w. 3 equal warts	in or by water	*Rumex hydrolapathum* Water Dock
	fr. w. 1 large wart & sometimes 2 small ones	1–2 m high, in a few waste places	*Rumex patientia* Patience Dock
Fr. w. 1 large wart & sometimes 2 small ones — fr.	not confined to woods		*Rumex sanguineus* Wood Dock

▶

Fr. w. 3 warts	1–2 m high, often in water	*Rumex hydrolapathum* Water Dock
	usu. under 1 m high	*Rumex conglomeratus* Clustered Dock

URTICACEAE Nettle Family

The two Nettles both sting, but the other species do not.
The flowers are small, green to pink, with a 4-lobed calyx and no petals.
There are either 4, 5 or no stamens.

Stinging plant; lvs. opposite	lower lvs. shorter than their stalks	catkins about 1 cm	*Urtica urens* Small Nettle
	lower lvs. longer than their stalks	catkins up to 4 cm	*Urtica dioica* Common Nettle
Lvs. alternate up to 6 mm [a]	fls. solitary, pink	forms low mats on damp walls etc.	*Soleirolia soleirolii* Mind-your-own-business
Lvs. alternate up to 7 cm; stems usu. reddish [b]	fls. in clusters	typically on walls, banks	*Parietaria judaica* Pellitory-of-the-wall

CANNABACEAE Hop

A rough climbing plant with opposite 3–5 lobed leaves.
The male and female flowers are on separate plants, the male in loose clusters, the female in a small cone.

Fls. greenish; stem 4-angled	lf. as above	fr. cone like, 2–5 cm [c]	*Humulus lupulus* Hop

ULMACEAE Elms

Trees whose leaves are asymmetrical at their base [a] & [b].
The flowers appear as tufts of reddish stamens in February or March, well before the leaves are out.
The fruit is a winged disc up to 2 cm across.
A number of minor species and hybrids also occur.

b

a

Most lvs. 4–10 cm, w. 8–14 pairs of veins	lvs. rough above [a]	*Ulmus procera* English Elm
	Lvs. smooth above	*Ulmus minor* Smooth Elm
	v. variable tree	
Lvs. 8–16 cm w. 15–20 pairs of veins [b]	trunk usu. divides low down	*Ulmus glabra* Wych Elm

MYRICACEAE Bog Myrtle

A deciduous shrub up to 2 m tall, with reddish twigs and alternate leaves which are aromatic if crushed.
The male and female red-tinged catkins are usually on separate plants.

male

female

flowering stems

Fls. reddish, often appearing before the lvs.	lvs. yellow-dotted, aromatic	moors, bogs	*Myrica gale* Bog Myrtle

PLATANACEAE Plane Family

A large tree with flaky bark.
The flowers & fruit are in long-stalked spherical clusters.

fruit

Lvs. alternate, lobed and/or sharply toothed	see also diag. on p. 101	planted	*Platanus × acerifolia* London Plane

137

138

BETULACEAE Birch Family

Trees and shrubs, bearing catkins, of which the male are drooping and longer than the female.
The fruiting catkins have many small 3-lobed (*Betula*) or 5-lobed (*Alnus*) scales.
The leaves are alternate.

Catkins in fl. before lvs. appear, but small	bark brown; lvs. rounded [a]	common in wet places	*Alnus glutinosa* Alder
	bark grey, smooth; lvs. pointed		*Alnus incana** Grey Alder
Low shrub w. rounded lvs. [b]		Scottish moors	*Betula nana** Dwarf Birch
Trunk white above, black & fissured below	twigs smooth; lvs. w. large & small teeth [c]		*Betula pendula* Silver Birch
Trunk grey/brown	twigs downy; lf. teeth mostly same size [d]		*Betula pubescens* Downy Birch

CORYLACEAE Hazel Family

Trees or shrubs, bearing catkins.
In Hornbeam both male and female catkins dangle.
In Hazel only the male catkins (Lamb's-tails) dangle; the female flowers look like tiny buds with 2 red styles.
The leaves are toothed and alternate.

Catkins yellow, opening before lvs.; lf. shape [e]	young lvs. downy below; twigs w. reddish hairs	fr. a nut	*Corylus avellana* Hazel
Catkins greenish, opening w. lvs.; lf. shape [f]	only veins hairy below; twigs downy	fr. w. 3-lobed wing [g]	*Carpinus betulus* Hornbeam

FAGACEAE Beech Family

Trees bearing flowers in catkins or clusters.
The fruit is a nut set in a green or a woody cup.
The leaves are alternate.

		a	b	c
Lvs. 10–25 cm, lanceolate toothed		catkins 12–20 cm long	nuts in a green spiny cup; [Horse-chestnut is on p. 000]	*Castanea sativa* Sweet Chestnut
Evergreen; lvs. dark above, downy below		young lvs. spiny; fr. an acorn	sometimes naturalized	*Quercus ilex* Evergreen Oak
Lvs. lobed NOTE: *Q. petraea* and *Q. robur* often hybridize	acorn cups w. long spreading scales [d]	lf. lobes usu. acute [c]		*Quercus cerris* Turkey Oak
	If. base tapering [b]	acorns almost stalkless		*Quercus petraea* Sessile Oak
	If. base lobed [a]	acorns long-stalked		*Quercus robur* Pedunculate Oak
Lvs. ovate, slightly wavy			nut 3-sided in a bristly husk	*Fagus sylvatica* Beech

139

SALICACEAE Poplars and Willows

Trees and shrubs. The leaves are nearly always alternate and often have stipules, though these soon fall in the Poplars.
The flowers are in catkins (the male and female on separate plants) which may open before the leaves.

The species in this family offer difficulties of identification not excelled by any other in the British Isles.
The B.S.B.I. Handbook *Willows and Poplars* provides up-to-date keys, illustrations, and descriptions, and includes many hybrids and cultivated varieties. But two warning notes are sounded: 'Poplar hybrids of inscrutable complexity are widely cultivated' and 'No Willow key yet devised will prove infallible'.
Another useful key to the larger lowland Willows has been produced by the AIDGAP project of The Field Studies Council.

Because of these difficulties it is not possible to produce here simple keys set out like the others in this book. However the Poplar keys should work with the majority of mature native trees growing wild; but the Willow lists aim only to give a lead towards identification, no hybrids being included.

Lvs. absent or v. immature	catkins dangling; trees	
	lf. buds w. several scales catkin [a]	Section 2 page 142 Poplars
	lf. buds w. only 1 scale	Section 4 page 145 Willows
	catkins more or less erect [b]	
	trees or shrubs	Section 4 page 145 Willows
Lvs. present	buds w. only 1 scale; trees & shrubs catkins more or less erect [b]	
	lvs. lanceolate or linear or ovate	Section 3 page 143 Willows
	buds w. several scales; trees	
	lvs. rounded or ovate/triangular or palmately lobed	Section 1 page 141 Poplars

SALICACEAE Section 1 Poplar in leaf

NOTE: Leaves on any one tree may vary considerably. Typical leaves from the tree, not from suckers, should be used for comparison with the illustrations.

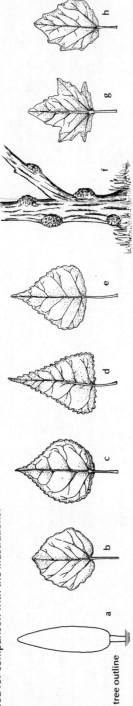

Remember: many cultivated varieties and hybrids also exist.

Lvs. hairless, rounded. w. flattened stalk	sucker lvs. ovate, downy below	lf. [b]	*Populus tremula* Aspen
	Young lvs. w. balsam smell	lf. [c]	*Populus × gileadensis** Balsam Poplar
Lvs. triangular, acute; no suckers	tree v. narrow in outline [a]	planted; lf. [d]	*Populus nigra 'Italica'* Lombardy Poplar
	young lvs. bronze; branches curve up	widely planted; lf. [e]	*Populus × canadensis* Italian Poplar
	branches spread and curve down	trunk w. swollen bosses [f]	*Populus nigra** Black Poplar
Some lvs. palmately lobed [g], densely downy white beneath			*Populus alba* White Poplar
Lvs. various, toothed but hardly lobed [h]			*Populus canescens* Grey Poplar

141

SALICACEAE Section 2 Poplars with catkins but no leaves.

Male and female catkins are on different trees.
The male flowers have 5 or more reddish anthers,
the female have 2 stigmas which are variously lobed.

Remember: many cultivated varieties and hybrids also exist.

Tree w. v. narrow outline [c]	catkins almost always male	planted	*Populus nigra 'Italica'* Lombardy Poplar
Catkins male only, w. 20–25 stamens per fl. [a]	young twigs slightly angled	widely planted	*Populus × canadensis* Italian Poplar
Catkins furry, especially the male	twigs hairless; buds slightly sticky	fl. bracts w. long teeth as well as hairs	*Populus tremula* Aspen
	twigs downy all over	catkins almost always female	*Populus alba* White Poplar
	twigs only downy near buds	stamens 8–15 per fl.	*Populus canescens* Grey Poplar
Twigs yellowish	seldom has suckers; trunk w. swollen bosses [d]		*Populus nigra** Black Poplar
Twigs brown; catkins female only	often has suckers	buds long, sticky, balsam-scented	*Populus × gileadensis** Balsam Poplar

SALICACEAE Section 3 Willows with well developed leaves

Remember: this section is not a key, only a list of probabilities

Low or creeping shrubs up to 150 cm high

Species	Description	Leaf
Salix herbacea Dwarf Willow	under 10 cm high; lvs blunt [e]; in North and on mountains	leaf [e]
Salix myrsinites Whortle-leaved Willow	up to 40 cm high Scottish mountains*	leaf [g]
Salix lapponum Downy Willow	mountain rocks in Cumbria and Scotland*	leaf [i]
Salix repens Creeping Willow	likes sandy or peaty soil; widespread	leaf [f]
Salix aurita Eared Willow	lvs. wrinkled; stipules large and conspicuous [b]; ridges under bark of 2–4 year old twigs [a]; widespread and common	leaf [l]
Salix purpurea Purple Willow	lvs. often in pairs	leaf [c]

Taller shrubs 1.5–5 m high

Species	Description	Leaf
Salix purpurea Purple Willow	lvs. often in pairs	leaf [c]
Salix pentandra Bay Willow	male fls. w. 5 or more stamens	leaf [j]
Salix triandra Almond Willow	male fls. w. 3 stamens compare p. 145 [a + b]	leaf [o]
Salix viminalis Osier	lvs. 10 cm or more long, linear; branches long, straight, flexible lowlands	leaf [d]
Salix nigricans Dark-leaved Willow	not in Southern England	leaf [n]
Salix phylicifolia Tea-leaved Willow	not S. of Lancs–Yorks	leaf [k]
Salix aurita Eared Willow	ridges under bark of 2–4 year old twigs [a]; widespread and common; stipules large & conspicuous, lvs. wrinkled	leaf [b, l]

a

b

c

d

e

f

g

▶

143

Salix cinerea Grey Willow	ridges under bark of 2–4 year old twigs [a] on p. 143; widespread and common	leaf [m]
Salix caprea Goat Willow	widespread and common	leaf [p]

Small trees 5–10 m high

Salix cinerea Grey Willow	ridges under bark of 2–4 year old twigs [a]; widespread and common	leaf [m]
Salix caprea Goat Willow	widespread and common	leaf [p]
Salix pentandra Bay Willow	male fls. w. 5 or more stamens; not common	leaf [j]
Salix alba White Willow	lvs. silvery, most at least 5 times as long as wide; widespread	leaf [h]
Salix fragilis Crack Willow	most lvs. at least 5 times as long as wide; twigs brittle; widespread	leaf [q]

Trees well over 10 m high

Salix alba White Willow	lvs. silvery, most at least 5 times as long as wide; widespread	leaf [h]
Salix fragilis Crack Willow	most lvs. at least 5 times as long as wide; twigs brittle; widespread	leaf [q]

h   i   j

k   l

m   n   o   p   q

SALICACEAE Section 4 Willows with catkins but no (or only very young) leaves

Remember: this section is not a key, only a list of probabilities

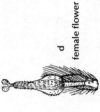

a
male catkin

b
male flower

c
female catkin

d
female flower

Low or creeping shrubs up to 150 cm high	
Salix repens Creeping Willow	likes sandy or peaty soil; widespread
Salix lapponum Downy Willow	mountain rocks in Cumbria and Scotland*
Salix purpurea Purple Willow	If. buds often in pairs; only 1 stamen to each male flower compare [b]
Salix aurita Eared Willow	ridges under bark of 2–4 year old twigs [e]; widespread and common
Taller shrubs 1.5–5 m high	
Salix purpurea Purple Willow	If. buds often in pairs; only 1 stamen to each male fl. style as [f]
Salix nigricans Dark-leaved Willow	not in Southern England usu. some lvs. w. catkins; female fl. [g]
Salix viminalis Osier	branches long, straight, flexible; lowlands; usu. planted; female fl. [h]
Salix aurita Eared Willow	ridges under bark of 2–4 year old twigs [e]; widespread and common
Salix cinerea Grey Willow	ridges under bark of 2–4 year old twigs [e]; widespread and common

e

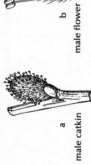

f

g

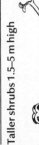

h

female flowers (ovary & stigmas)

▶

i	widespread and common; female fl. [i]	
Small trees 5–10 m high		*Salix caprea* Goat Willow
	ridges under bark of 2–4 year old twigs [e] on p. 145; widespread and common, but normally a shrub	*Salix cinerea* Grey Willow
	widespread and common	*Salix caprea* Goat Willow

ERICACEAE Heath Family

Flowers with 4 or 5 lobes, often with a bell-shaped corolla.
In *Rhododendron* the corolla is more funnel-shaped and the lobes large.
There are usually either 8 or 10 stamens. The fruit may be a berry or a capsule.
This is a family of shrubs (mostly low ones) typically growing on heaths, moors and mountains.

Petals almost free	sepals & petals all alike [c]	lvs. up to 2mm, overlapping, in 4 rows	*Calluna vulgaris* Heather
	petals bent backwards [d]; fr. a berry	creeping bog plant; lvs. alternate	*Vaccinium oxycoccos** Cranberry

Lvs. in whorls	stamens protrude [b]	Cornwall	*Erica vagans** Cornish Heath
	lvs. on flowering stems usu. 3 to a whorl	lvs. fringed w. long hairs [e]; S.W. only	*Erica ciliaris* Dorset Heath
		calyx hairless [a] leaves as [f]	*Erica cinerea* Bell Heather
	lvs. on flowering stems 4 to a whorl [g]		*Erica tetralix* Cross-leaved Heath

Lvs. up to 2 mm, overlapping in 4 rows		flower [c]	*Calluna vulgaris* Heather
Lvs. opposite; corolla 5-lobed; stamens 5	Scottish mts.		*Loiseleuria procumbens** Trailing Azalea
Lvs. 5–12 cm long	Lf. base rounded; fls. white/pink	up to 1 m high	*Gaultheria shallon* Shallon
	Lf. base tapering	up to 3 m high	*Rhododendron ponticum* Rhododendron
Lvs. nearly linear; bog plant	fr. a dry capsule		*Andromeda polifolia** Bog Rosemary
Lvs. finely or slightly toothed	lvs. dotted w. glands beneath	cor. white/pink; berry red; leaf [a]	*Vaccinium vitis-idaea* Cowberry
	fls. usu. in ones, pinkish; berry blackish purple	widespread on acid soils; leaf [b]	*Vaccinium myrtillus* Bilberry
	cor. whitish; berry black	N. Scotland only; leaf [c]	*Arctostaphylos alpinus** Alpine Bearberry
Lvs. entire	fls. in clusters of 5–12; berry red	evergreen; not in S.	*Arctostaphylos uva-ursi* Bearberry
	fls. in clusters of 1–4; berry black	deciduous; not in S.	*Vaccinium uliginosum** Bog Bilberry

PYROLACEAE Wintergreens

The Wintergreens have spikes of whitish flowers.
There are 5 sepals and petals, and 10 stamens.
The stigma is 5-lobed.
The evergreen leaves are mostly radical.

Pyrola minor

Inflor. one-sided	style usu. straight, protruding	N. & W.	*Orthilia secunda** Serrated Wintergreen
Style straight or nearly so	style shorter than stamens		*Pyrola minor* Common Wintergreen
	style longer than stamens	N.	*Pyrola media** Intermediate Wintergreen
Style curved, protruding			*Pyrola rotundifolia** Round-leaved Wintergreen

MONOTROPACEAE Bird's-nest

Yellowish plants without green leaves.
The flowers, usually in a drooping spike, have 4 or 5 sepals and petals and 8–10 stamens.
Do not confuse this plant with the similar looking Bird's-nest Orchid, p. 211, whose flowers have a 2-lobed lip.

Sepals strap-shaped	lvs. bract-like, 5–10 mm, yellowish	woods and dunes; up to 30 cm high	*Monotropa hypopitys** Yellow Bird's-nest

EMPETRACEAE Crowberry

A small heather-like shrub of high moor and mountain.
The perianth has 6 segments. There are either 3 or no stamens.
The leaves are small, narrow, and alternate or crowded.

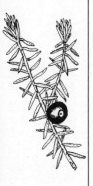

Fls. pinkish, about 2 mm across	fr. black	*Empetrum nigrum* Crowberry

PLUMBAGINACEAE Sea-lavender Family

Flowers with a 5-lobed calyx and 5 petals, stamens and styles.
The leaves are simple and all radical.

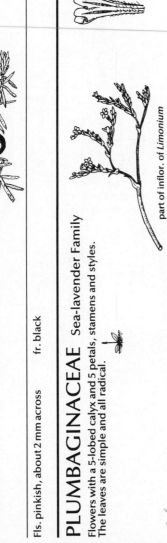

part of inflor. of *Limonium*

Calyces

Lvs. linear	fls. pinkish, in round heads	coasts & mts.	*Armeria maritima* Thrift
Calyx w. 5 entire blunt lobes [a]	lvs. faintly 3-veined	cliffs & shingle	*Limonium binervosum* Rock Sea-lavender
Branches start below half-way; calyx [b]	lf. veins obscure; inflor. loose	saltmarshes	*Limonium humile** Lax-flowered Sea-lavender
Branches from above half-way; calyx [c]	lf. veins distinct; fls. closely packed	saltmarshes	*Limonium vulgare* Common Sea-lavender

PRIMULACEAE Primrose Family

Most flowers in this family have a 5-lobed calyx (or 5 sepals) and 5 petals which often spring from a funnel-shaped tube. *Glaux* has no petals but coloured sepals. There are normally 5 stamens and one style.

Aquatic w. submerged pinnate lvs.	fls. lilac			*Hottonia palustris** Water-violet
No stem lvs.	fls. purplish; lvs. mealy beneath	sepals acute; N. England & S. Scotland ∧		*Primula farinosa* Bird's-eye Primrose
		sepals blunt; N. Scotland only ∩		*Primula scotica** Scottish Primrose
	fls. solitary, 25 mm or more across			*Primula vulgaris* Primrose
	inflor. usu. 1-sided	corolla tube w. folds or ridges; widespread		*Primula veris* Cowslip
		corolla tube not folded; E. Anglia		*Primula elatior** Oxlip
	inflor. spreading; calyx strongly ridged			*Primula veris × vulgaris* False Oxlip
Fls. yellow	low, creeping plant	sepals narrowly acute		*Lysimachia nemorum* Yellow Pimpernel
		sepals broadly ovate		*Lysimachia nummularia* Creeping-Jenny
	inflor. extends to top of plant	lvs. 2–4 together		*Lysimachia vulgaris* Yellow Loosestrife
	fl. clusters axillary; below top lvs.	lvs. usu. in pairs		*Lysimachia thyrsiflora** Tufted Loosestrife

Fls. 5 mm or less, stalkless, in lf. axils	sepals pink; petals absent		*Glaux maritima* Sea-milkwort
	petals under 1 mm, shorter than sepals [a]	usu. under 5 cm high [c]	*Anagallis minima** Chaffweed
Fls. white	petals 5; lvs. alternate	wet places	*Samolus valerandi* brookweed
	petals 5–9; most lvs. in a whorl	northern woods & moors	*Trientalis europaea** Chickweed Wintergreen
Creeping bog plant	petals pink, veined		*Anagallis tenella* Bog Pimpernel
Fls. usu. red or pink, rarely blue			*Anagallis arvensis* Scarlet Pimpernel

BUDDLEJACEAE Buddleja

The flowers, with 4-lobed corollas, are in dense inflorescences, much favoured by butterflies.
The leaves are green above and whitish below.

Inflor. purplish, 10–30 cm long	lvs. 10–25 cm	garden escape often naturalized	*Buddleja davidii* Butterfly-bush

OLEACEAE Olive Family

Trees or shrubs with opposite leaves.
The flowers, except Ash, have 4 petals and 2 stamens, and appear in clusters.

Tree; lvs. pinnate; buds black	fls. (tufts of stamens) appear before lvs.	*Fraxinus excelsior* Ash
Lvs. 3–6 cm; fls. white; fr. a black berry	lvs. lanceolate young twigs downy	*Ligustrum vulgare* Wild Privet
	lvs. oval; young twigs hairless	*Ligustrum ovalifolium* Garden Privet
Lvs. 5–12 cm; fls. lilac or white	garden escape	*Syringa vulgaris* Lilac

APOCYNACEAE Periwinkles

Creeping shrubs with opposite evergreen leaves.
The flowers are large, bluish or white, with 5 sepals and petals.

Fls. 40–50 mm across; lvs. w. fine hairs	sepals linear w. fine hairs	*Vinca major* Greater Periwinkle
Fls. 25–30 mm across; lvs. hairless	sepals lanceolate hairless	*Vinca minor* Lesser Periwinkle

GENTIANACEAE Gentian Family

Gentians have a tubular corolla with from 4 to 8 lobes (petals) and one stamen between each lobe. There are the same number of sepals (or calyx lobes) as petals. The leaves are basal or opposite, simple and untoothed.

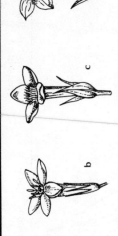
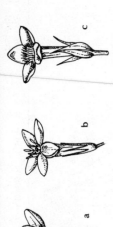

a | b | c | d

Fls. yellow	sepals & petals 6–8	calcareous grassland & dunes	*Blackstonia perfoliata* Yellow-wort
	sepals & petals 4; lvs. up to 6 mm, linear	up to 12 cm high; S. & W. [a]	*Cicendia filiformis** Yellow Centaury
Fls. pink; sepals linear, keeled	most fls. stalked; inflor. loose; no basal rosette of lvs.		*Centaurium pulchellum* Lesser Centaury
	basal lvs. up to 5 mm wide w. 3 faint veins	coasts, mainly N. & W.	*Centaurium littorale** Seaside Centaury
	basal lvs. over 5 mm wide w. 3–7 clear veins	widespread [b]	*Centaurium erythraea* Common Centaury
Corolla purplish w. fringe round its throat	calyx w. 2 large & 2 small lobes	[c]	*Gentianella campestris* Field Gentian
	in fl. Aug.–Oct.	calyx only ½ length of corolla tube	*Gentianella germanica** Chiltern Gentian
		calyx nearly as long as corolla tube	*Gentianella amarella* Autumn Gentian
	in fl. May–June	calcareous grassland in S.	*Gentianella anglica** Early Gentian
Corolla w. 5 large & 5 small lobes [d]	wet heaths		*Gentiana pneumonanthe** Marsh Gentian

MENYANTHACEAE Bogbean Family

Aquatic or bog plants with 5 fringed petals.
The leaves are alternate.

Lvs. rounded, floating (like small Water-lily lvs.)	fls. yellow, about 3 cm across [a]	*Nymphoides peltata** Fringed Water-lily
Lvs. w. 3 lflets rising above surface	fls. white/pink w. white hairy fringes [b]	water or bog plant	*Menyanthes trifoliata* Bogbean

POLEMONIACEAE Jacob's-ladder

Flowers with 5 sepals, petals and stamens.
The leaves are alternate and pinnate, the lower ones stalked, the upper hardly so.

Fls. blue or white, bell-shaped 2–3 cm across	on limestone in N.	*Polemonium caeruleum** Jacob's ladder

BORAGINACEAE Borage Family

Flowers with 5 sepals, petals and stamens.
Inside the ripe calyx can be seen 4 nutlets, each containing 1 seed.
The leaves are alternate and simple, often downy or rough.

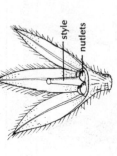

style

nutlets

Lvs. hairless, bluish, fleshy, entire	coastal shingle, in N. only	fls. pink becoming blue	*Mertensia maritima** Oysterplant

Fls. white or cream	corolla under 2 mm across becoming blue		*Myosotis discolor* Changing Forget-me-not
	lvs. w. obvious side veins [a]; fr. white	calyx lobes nearly as long as corolla tube	*Lithospermum officinale* Common Gromwell
	no side veins on lvs. [b]; fr. brown		*Lithospermum arvense* Field Gromwell
	calyx lobes longer than calyx tube	lower lvs. the largest; stem well branched	*Symphytum officinale* Common Comfrey
		middle lvs. longer than lower ones	*Symphytum tuberosum* Tuberous Comfrey
	calyx lobes shorter than calyx tube		*Symphytum orientale* White Comfrey
Some stamens protrude well beyond corolla	stamens protrude in a close column	fl. shape	*Borago officinalis* Borage
	stamens protrude loosely		*Echium vulgare* Viper's-bugloss
Some fls. pink, red, purple or maroon	30–90 cm high; smells of mice; corolla maroon	fl. shape [c]	*Cynoglossum officinale* Hound's-tongue
	sepals long, narrow; lvs. linear lanceolate	fls. purplish blue; calcareous soil; fl. shape [e]	*Lithospermum purpurocaeruleum** Purple Gromwell*
	stem 30–120 cm high w. stiff hairs	fls. pink, becoming blue; fl. shape [d]	*Symphytum × uplandicum* Russian Comfrey
		fls. purplish (or creamy white); fl. shape [d]	*Symphytum officinale* Common Comfrey

▶

Erect plant, unpleasantly rough to handle	lvs. white-spotted over 10 cm; corolla pink becoming blue; corolla tube at least twice as long as lobes	radical lvs. w. tapering base; Hants & Dorset	*Pulmonaria longifolia** Narrow-leaved Lungwort
		radical lvs. ovate, stalked	*Pulmonaria officinalis* Lungwort
	corolla bent [b]; lvs. wavy, bristly	fls. blue	*Anchusa arvensis* Bugloss
	corolla tube longer than lobes [a]	fls. pink, becoming blue	*Symphytum × uplandicum* Russian Comfrey
	corolla tube shorter than lobes	fls. blue; calyx and nutlets as [c]	*Pentaglottis sempervirens* Green Alkanet
Corolla tube at least twice as long as corolla lobes; fls. pink/blue/purple	calyx divided almost to the base [d]	calcareous places; not in N.	*Lithospermum purpurocaeruleum* Purple Gromwell*
	radical lvs. lanceolate w. tapering base	Hants & Dorset	*Pulmonaria longifolia** Narrow-leaved Lungwort
	radical lvs. ovate, stalked		*Pulmonaria officinalis* Lungwort
Hairs on calyx tube spreading, not appressed	fls. up to 3 mm across	fls. cream at first, turning blue	*Myosotis discolor* Changing Forget-me-not
		all fls. blue	*Myosotis ramosissima* Early Forget-me-not
	fls. concave, up to 5 mm across	style shorter than calyx tube [e]	*Myosotis arvensis* Field Forget-me-not
	fls. flat 6–8 mm across	style longer than calyx tube [f]	*Myosotis sylvatica* Wood Forget-me-not
Style at least as long as calyx tube	lvs. only about twice as long as wide	fls. pale blue [g]; in N.	*Myosotic stolonifera** Pale Forget-me-not
	calyx teeth as long as calyx tube [h]	fr. stalk 3–5 times as long as calyx	*Myosotic secunda* Creeping Forget-me-not

	calyx teeth shorter than tube [a]; fls. 2–5 mm across	*Myosotis scorpioides* Water Forget-me-not
Style often only about ½ length of calyx tube [b] or [c]	fls. 2–5 mm across	*Myosotis laxa* Tufted Forget-me-not

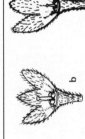
a

b

c

CONVOLVULACEAE

Flowers with a funnel-shaped corolla, sometimes 5-lobed.
There are 5 sepals, 5 stamaens and 2 stigmas.
These are all creeping or climbing plants with alternate leaves,
except the leafless parasite Dodder.

Dodder flowers

d

e

f

Lf less red-stemmed parasite; fls. up to 5 mm in dense heads up to 15 mm across [d]	style longer than ovary; stamens protrude	*Cuscuta epithymum* Dodder
	on Gorse, Heather, Thyme, etc.	
	style shorter than ovary; stamens inside corolla	*Cuscuta europaea** Greater Dodder
	usu. on Nettle or Hop	
2 large bracts round calyx	creeping seaside plant; lvs. roundish	*Calystegia soldanella* Sea Bindweed
	in sand or shingle; fls. pink	
	fl. bracts inflated, overlapping each other [f]	*Calystegia silvatica* Large Bindweed
	fls. white, 5–7 cm across	
	fl. bracts hardly overlapping	*Calystegia sepium* Hedge Bindweed
	fls. white, usu. 3–5 cm across	
No bracts round calyx		*Convolvulus arvensis* Field Bindweed
Fls. about 3 cm across [e], white or pink		

SOLANACEAE

Flowers with a bell- or funnel-shaped 5-lobed corolla from which the stamens usually project.
The fruit is a berry or capsule, often poisonous.
The leaves are alternate

Corolla funnel-like, 6–8 cm across [a]	fls. white or purple		fr. 4–5 cm, spiny	*Datura stramonium** Thorn-apple
Stamens protrude in a tight yellow cone	corolla white		ripe fr. a black berry w. small calyx	*Solanum nigrum* Black Nightshade
			fr. a green berry w. large calyx	*Solanum sarrachoides** Green Nightshade
	corolla purple		fr. green to red	*Solanum dulcamara* Bittersweet
Fls. 2–3 cm long or across	corolla yellow w. purple veins; fr. a capsule		plant downy/sticky; a very poisonous plant	*Hyoscyamus niger* Henbane
	corolla purple [b]; fr. a black berry		a very poisonous plant	*Atropa belladonna* Deadly Nightshade
Corolla lobes shorter than tube	lvs. narrowly lanceolate; a shrub		fr. red, ovoid	*Lycium barbarum* Duke of Argyll's Teaplant
Corolla lobes not shorter than tube	lvs. almost ovate; a shrub		fr. red, ovoid	*Lycium chinense* China Teaplant

158

SCROPHULARIACEAE Figwort Family

A family with a great variety of flower form and colour.
The calyx and corolla have 4 or 5 lobes.
There are 2 stamens in *Veronica*, 5 in *Verbascum* and 4 in the other genera.
The fruit is a many-seeded capsule, not 4 nutlets as in the Labiates, whose flowers are sometimes similar.

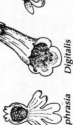   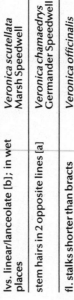 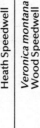 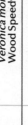

| *Veronica* | *Verbascum* | *Linaria* | *Scrophularia* | *Mimulus* | *Pedicularis* | *Melampyrum* | *Euphrasia* | *Digitalis* | *Cymbalaria* |

Fls. w. 2 stamens and a 4-lobed corolla (*Veronica*) Lvs. opposite	Section 1 page 159
Fls. w. 4 or 5 stamens	Section 2 page 161

SCROPHULARIACEAE Section 1 Flowers with 2 stamens and 4-lobed corolla

Fls. on almost lfless spikes which rise from lf. axils	stem hairy	lvs. linear/lanceolate [b]; in wet places	*Veronica scutellata* Marsh Speedwell
		stem hairs in 2 opposite lines [a]	*Veronica chamaedrys* Germander Speedwell
		fl. stalks shorter than bracts lf. [c]	*Veronica officinalis* Heath Speedwell
		lvs. w. distinct stalk [d]	*Veronica montana* Wood Speedwell

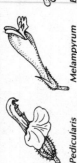 

a b c d

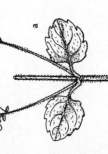

▶

fl. spikes in opposite pairs	lvs. w. short stalks [a]		*Veronica beccabunga* Brooklime
		fls. pink lf [b]	*Veronica catenata* Pink Water-speedwell
		fls. blue lf. [b]	*Veronica anagallis-aquatica* Blue Water-speedwell
fl. spikes alternate; up to 10 fls. in each spike	lvs. narrow; in wet places [c]		*Veronica scutellata* Marsh Speedwell
Fls. in dense terminal spike	lf. [d]		*Veronica spicata** Spiked Speedwell
Upper lvs. w. v. narrow lobes	E. Anglia		*Veronica triphyllos** Fingered Speedwell
Fls. almost stalkless	lvs. entire, hairless [e]		*Veronica serpyllifolia* Thyme-leaved Speedwell
	sepals longer than petals or fr. lf. [f]		*Veronica arvensis* Wall Speedwell
	petals longer than sepals or fr.	mts. in Scotland	*Veronica alpina** Alpine Speedwell
Fls. pale lilac; sepals cordate	lvs. often ivy-shaped		*Veronica hederifolia* Ivy-leaved Speedwell

a   b   c   d   e   f

Fl. stalks 2–4 times as long as lvs.			*Veronica filiformis* Slender Speedwell
Fl. stalks longer than lvs. (up to twice as long)	fls. 8–12 mm across		*Veronica persica* Common Field-speedwell
	fls. 3–6 mm across	Corolla usu. all blue	*Veronica polita* Grey Field-speedwell
		Corolla w. one or more white lobes	*Veronica agrestis* Green Field-speedwell

SCROPHULARIACEAE   Section 2   Flowers with 4 or 5 stamens

Fls. w. 5 often hairy stamens; tall plants; inflor. a spike which may be branched	stamen hairs white	only 3 stamens v. hairy; fls. yellow	*Verbascum thapsus* Great Mullein
		all stamens hairy; fls. usu. white	*Verbascum lychnitis** White Mullein
fls. yellow or white	lvs. hairless	fl. stalks longer than calyx; fls. in ones	*Verbascum blattaria** Moth Mullein
		fl. stalks shorter than calyx; fls. 1–5 together	*Verbascum virgatum* Twiggy Mullein
	lvs. hairy	5–10 fls. to each bract	*Verbascum nigrum* Dark Mullein
Corolla w. a spur at its base	lvs. ivy-shaped [a]; fls. lilac		*Cymbalaria muralis* Ivy-leaved Toadflax
	lvs. ovate w. a few teeth [b]		*Kickxia spuria* Round-leaved Fluellen

	many lvs. triangular w. 2 points at base [a]		_Kickxia elatine_ Sharp-leaved Fluellen
	fls. yellow w. orange spot		_Linaria vulgaris_ Common Toadflax
	fls. in narrow upright spike	fls. bright to dark purple	_Linaria purpurea_ Purple Toadflax
		fls. pale violet, veined	_Linaria repens_ Pale Toadflax
	fls. in ones, purplish		_Chaenorhinum minus_ Small Toadflax
Small creeping plant of damp places; corolla 5-lobed, up to 3 mm across	lvs. narrow, entire; fls. whitish		_Limosella aquatica_* Mudwort
	lvs. round, lobed; fls. yellow & pink	S. & W. only	_Sibthorpia europaea_* Cornish Moneywort (lf.)
Upper lvs. alternate	plant 50–150 cm tall		_Digitalis purpurea_ Foxglove
	lvs. entire	fls. over 30 mm	_Antirrhinum majus_ Snapdragon
		fls. under 20 mm [b]	_Misopates orontium_ Lesser Snapdragon
	lvs. pinnate w. toothed lflets	calyx hairy; upright plant w. few branches fl. [c]	_Pedicularis palustris_ Marsh Lousewort
		calyx hairless; plant usu. well branches from base	_Pedicularis sylvatica_ Lousewort
	lvs. small, toothed [e], lower ones opposite	fls. white, lilac veined [d]	_Euphrasia officinalis_ (agg.) Eyebright

Fls. purplish brown	lvs. downy	in S.W. only	*Scrophularia scorodonia** Balm-leaved Figwort
	stem winged on all 4 angles; sepals w. broad white border	scale inside corolla upper lip 2-lobed [a]	*Scrophularia umbrosa** Green Figwort
		scale inside corolla rounded [b]; common	*Scrophularia auriculata* Water Figwort
	stem square but not winged		*Scrophularia nodosa* Common Figwort
Flks. white, veined w. violet	lf. shape	sepals hardly bordered	*Euphrasia officinalis* (agg.) Eyebright
Fls. pink	lvs. lanceolate; w. a few teeth		*Odontites verna* Red Bartsia
Lvs. linear/lanceolate almost entire; fls. pale to bright yellow	fls. mostly in pairs	upper lvs. toothed at base; fls. 12–22 mm	*Melampyrum pratense* Common Cow-wheat
		fls. under 12 mm long; in N. only	*Melampyrum sylvaticum** Small Cow-wheat
	fls. in short, dense, terminal spike	E. Anglia	*Melampyrum cristatum** Crested Cow-wheat
Fls. w. hooded corolla, yellow	calyx v. inflated in fr. [c]		*Rhinanthus minor* (agg.) Yellow Rattle
	plant sticky fl. [d]	main S. & W.	*Parentucellia viscosa* Yellow Bartsia

164

Corolla lobes spreading, flat, yellow	plant sticky/hairy		*Mimulus moschatus* Musk
	corolla w. large red spots and open throat		*Mimulus luteus* Blood-drop-emlets
plants of wet places	corolla w. small spots and almost closed throat	fl. [a]	*Mimulus guttatus* Monkeyflower
Fls. globular, w. tiny lobes, yellow	30–80 cm high; lvs. broad, deeply toothed		*Scrophularia vernalis* * Yellow Figwort

OROBANCHACEAE Broomrape Family

Brownish plants without leaves, only thick scales on the stem.
They are parasites on the roots of other plants.
The flowers, borne in a spike, are 2-lipped and variously lobed.
There are 4 stamens.
There are also several much rarer species which parasitize one particular plant. Identification is difficult as the differences are small and it is often not possible to be sure of the host plant.

Lathraea   *Orobanche*

Calyx w. 4 short equal lobes fl. shape [b]	inflor. usu. 1-sided	often on Hazel or Elm roots	*Lathraea squamaria* Toothwort
Stamens joined to corolla 2–6 mm above base; fl. shape [c]	inflor. dense; filaments hairy below	on Knapweed	*Orobanche elatior* * Knapweed Broomrape
	inflor. loose; young stigmas yellow	on Ivy	*Orobanche hederae* * Ivy Broomrape
	stigmas purple; corolla 10–16 mm	on various plants, especially Leguminosae	*Orobanche minor* Common Broomrape
Stamens arise from base of corolla tube	corolla 20–25 mm	usu. on Broom or Gorse	*Orobanche rapum-genistae* * Greater Broomrape

LENTIBULARIACEAE Butterwort Family

Insectivorous aquatic and bog plants, with zygomorphic yellow or purplish flowers. Each flower has 5 sepals and a 2-lipped or 5-lobed corolla, with a spur. Butterwort leaves are simple and untoothed in a basal rosette. Bladderworts have finely divided submerged leaves among which may be found the tiny bladders in which aquatic animacules are trapped.

Aquatic plant w. finely divided lvs. and small bladders; fls. yellow [a]	bladders only on lfless branches	lf. segments w. tiny bristles along edges	*Utricularia intermedia** Intermediate Bladderwort
		lf. segments entire	*Utricularia minor* Lesser Bladderwort
	bladders only on lfy branches; lf. segments w. bristles along edges	edges of lower corolla lip bent right down [a]	*Utricularia vulgaris* Greater Bladderwort
		lower corolla lip only wavy	*Utricularia australis** Western Bladderwort
	bladders on lfy & lfless branches	lf. segments entire	*Utricularia minor* Lesser Bladderwort
Bog plant w. rosette of entire lvs.; fls. in ones, long-staled, spurred [b]	fls. pale lilac, 7–9 mm	S.W. & N.W.	*Pinguicula lusitanica** Pale Butterwort
	fls. violet, 15–22 mm [b]	more widespread	*Pinguicula vulgaris* Common Butterwort

VERBENACEAE Vervain

The corolla is slightly 2-lipped somewhat resembling a mint, but with 5 lobes. There are 4 stamens and 4 nutlets inside the ripe calyx. The leaves are opposite and lobed.

Fls. pale lilac in narrow spikes	30–60 cm high	*Verbena officinalis* Vervain

165

LABIATAE Labiate Family

A large family with zygomorphic flowers very often in whorls. The flowers are normally 2-lipped, though not clearly so in *Mentha* or *Lycopus*.
The calyx is 5-toothed, and inside, when it is ripe, can be seen 4 nutlets.
There are 4 stamens, except in *Salvia* and *Lycopus* which usually have two.
The leaves are opposite and the stems more or less square.
On page 257 there is a Simplified Key to the commoner Labiates.

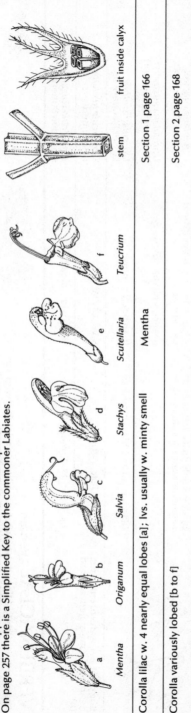

| Mentha | Origanum | Salvia | Stachys | Scutellaria | Teucrium | stem | fruit inside calyx |
| | a | b | c | d | e | f | |

Corolla lilac w. 4 nearly equal lobes [a]; lvs. usually w. minty smell	Mentha		Section 1 page 166
Corolla variously lobed [b to f]			Section 2 page 168

LABIATAE Section 1 Corolla lilac with 4 nearly equal lobes. Genus *Mentha*.

Mints vary considerably and hybridize freely, so that it is often difficult to assign an exact name to any particular specimen.
Five species and five hybrids are keyed out here.
Marjoram (*Origanum*) resembles the Mints but the corolla is more clearly 2-lipped. See diags.

		Mentha	Origanum
Semi-prostrate; lvs. under 12 mm wide	2 calyx teeth narrower than other 3		*Mentha pulegium** Pennyroyal
Fl. whorls all axillary; no terminal head of fls.	stamens inside corolla tube	calyx hairy all over	*Mentha* × *verticillata* Whorled Mint

Most fls. in a rounded terminal head (not a spike)	calyx hairy all over; stamens protrude from corolla tube	calyx hairy only at top	*Mentha × gentilis* * Bushy Mint
		calyx teeth approx. equilateral	*Mentha arvensis* Corn Mint
		calyx teeth longer than broad	*Mentha × verticillata* Whorled Miant
	calyx hairy only at top		*Mentha × smithiana* * Tall Mint
Stem & lvs. almost hairless	stamens protrude		*Mentha aquatica* Water Mint
	stamens inside corolla [a]; lvs. stalked		*Mentha × piperita* Peppermint
	stamens protrude; [b] lvs. not stalked		*Mentha spicata* Spear Mint
Lvs. hairy, wrinkled	stamens protrude from corolla	lvs. usu. obtuse; [c] mainly S. & W.	*Mentha suaveolens* Round-leaved Mint
		lvs. lanceolate; calyx tube hairless fl. [b]	*Mentha spicata* Spear Mint
		lvs. lanceolate; calyx tube hairy	*Mentha longifolia* Horse Mint
	stamens inside corolla tube		*Mentha × villosa* Large Apple-mint

a   b

c

LABIATAE Section 2 Corolla variously lobed (see page 166)

See also Simplified Key on page 257.

Lower corolla lip w. 2 humps near its base	corolla pale yellow, lower lip violet	corolla tube 22–34 mm long → *Galeopsis speciosa* Large-flowered Hemp-nettle
		corolla tube 13–20 mm → *Galeopsis tetrahit* Common Hemp-nettle
	stem w. stiff hairs and swollen nodes	*Galeopsis tetrahit* Common Hemp-nettle
	stem softly hairy	*Galeopsis angustifolia* Red Hemp-nettle
Fls. yellowish	lvs. w. 3 linear lobes; fl. [a] in S.E. leaf	*Ajuga chamaepitys** Ground-pine
a b c	corolla bright yellow; upper lip hooded [b]	*Lamiastrum galeobdolon* Yellow Archangel
	corolla pale green/yellow; no upper lip [c]	*Teucrium scorodonia* Wood Sage
Fls. mainly white	corolla 25–40 mm long, pink & white [a] in S.W.	*Melittis melissophyllum** Bastard Balm
a b c	upper lvs. almost pinnate calcareous grassland in S.	*Prunella laciniata** Cut-leaved Selfheal
	stamens only 2, protruding [b]; lvs. w. large teeth	*Lycopus europaeus* Gipsywort
	calyx 10-toothed; stamens inside corolla tube [c]	*Marrubium vulgare* White Horehound

			Species
	corolla purple-spotted; lvs. downy [d]		*Nepeta cataria* Cat-mint
	calyx w. 5 almost equal teeth [e]		*Lamium album* White Dead-nettle
	calyx w. 3 small & 2 larger teeth [f]	lvs. lemon-scented; garden escape	*Melissa officinalis* Balm
Corolla w/out upper lip	corolla pale green/yellow [g]		*Teucrium scorodonia* Wood Sage
	corolla usu. deep blue [h]		*Ajuga reptans* Bugle
Lvs. w. 3–7 sharp lobes	corolla pinkish, spotted		*Leonorus cardiaca** Motherwort
Stamens 0 or 2	corolla up to twice as long as calyx	calyx w. white hairs	*Salvia verbenaca* Wild Clary
	corolla up to 3 times as long as calyx	calyx only downy & glandular	*Salvia pratensis* Meadow Clary
Calyx w. only 2 lips; fls. usu. in pairs	fls. blue; plant up to 50 cm		*Scutellaria galericulata* Skullcap
	fls. pink/purple; usu. under 15 cm high		*Scutellaria minor* Lesser Skullcap
At least 2 stamens longer than upper lip of corolla	calyx w. 5 almost equal teeth	upright plant, 30–60 cm high	*Origanum vulgare* Marjoram
	stem under fl. head hairy on 2 opposite sites		*Thymus praecox* Wild Thyme
	fl. stems hairy along angles		*Thymus pulegioides* Large Thyme

▶

Calyx w. 3 short and 2 much longer teeth	corolla hooded; fls. in oblong terminal head	*Prunella vulgaris* Selfheal
	fls. in dense whorls mixed w. many bracts [b]	*Clinopodium vulgare* Wild Basil
	fl. clusters on axillary stalks, not in whorls on main stem	larger lvs. 2–5 cm; calyx teeth w. long hairs [a] → *Calamintha sylvatica* Common Calamint
		ripe calyx w. hairs protruding from inside → *Calamintha nepeta* Lesser Calamint
	calyx w. swollen base	*Acinos arvensis* Basil Thyme
Style longer than corolla tube		*Origanum vulgare* Marjoram
Inner 2 stamens longer than outer 2	beware: outer stamens may curve down and look shorter than inner ones	*Glechoma hederacea* Ground-ivy
Lvs. w. white blotch	garden escape	*Lamium maculatum* Spotted Dead-nettle
Upper lvs. rounded, stalkless half joined in pairs	[c, d]	*Lamium amplexicaule* Henbit Dead-nettle
Lower corolla lip w. 2 large lobes	lvs. sharply toothed; upper lf. stalks winged	*Lamium hybridum* Cut-leaved Dead-nettle

a

b

lf. shape

flower

d

leaf

c

		corolla tube longer than calyx	*Lamium purpureum* Red Dead-nettle
Most fls. packed in a dense terminal head		corolla reddish purple	*Stachys officinalis* Betony
Plant 30–100 cm high; fls. 12–18 mm long	smell strong, unpleasant; lvs. w. medium to long sitalks	fls. dark purple, mostly in spike above lvs.; calyx [a, b]	*Stachys sylvatica* Hedge Woundwort
		fls. pale purple, in whorls among lvs.; calyx [c]	*Ballota nigra* Black Horehound
	smell faint; lvs. almost stalkless		*Stachys palustris* Marsh Woundwort
Plant up to 30 cm	fls. 6–7 mm, pale purple	lvs. ovate, toothed	*Stachys arvensis* Field Woundwort
	fls. 15–20 mm, blue/violet	lvs. rounded, bluntly toothed	*Glechoma hederacea* Ground-ivy

in fl. in fr.
a b c

PLANTAGINACEAE Plantains

Except for Shoreweed (*Littorella*) the flowers are in dense spikes. The calyx and corolla are tiny but the long-stalked stamens (4 to each flower) make the inflorescence conspicuous.

Other plants with a similar appearance are:
Mousetail (5–10 stamens all at base of inflorescence) see p. 000.
Arrowgrass (6 stalkless stamens to each flower) see p. 000.
Adder's tongue (no stamens and only one leaf) see p. 000.

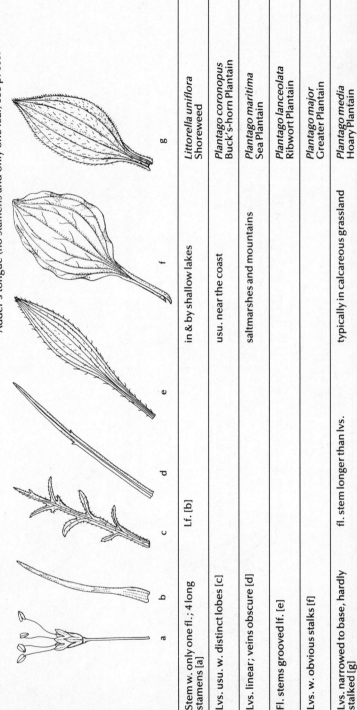

a   b   c   d   e   f   g

Stem w. only one fl.; 4 long stamens [a]	Lf. [b]	in & by shallow lakes	*Littorella uniflora* Shoreweed
Lvs. usu. w. distinct lobes [c]		usu. near the coast	*Plantago coronopus* Buck's-horn Plantain
Lvs. linear; veins obscure [d]		saltmarshes and mountains	*Plantago maritima* Sea Plantain
Fl. stems grooved lf. [e]			*Plantago lanceolata* Ribwort Plantain
Lvs. w. obvious stalks [f]			*Plantago major* Greater Plantain
Lvs. narrowed to base, hardly stalked [g]	fl. stem longer than lvs.	typically in calcareous grassland	*Plantago media* Hoary Plantain

CAMPANULACEAE Bellflower Family

Flowers with a tubular 5-lobed corolla, usually blue or blue/purple.
There are 5 calyx teeth, 5 stamens, and 2 or 3 stigmas on a single style.
The leaves are alternate and simple.

NOTE: A similar looking plant to Rampion and Sheep's-bit is Scabious,
but this has opposite leaves.

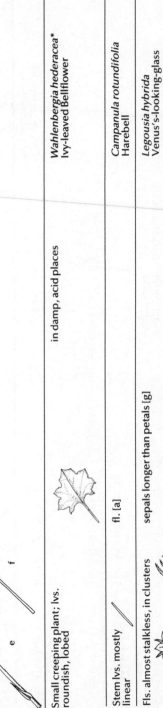

Fls. many, small, in a roundish head	young petals joined at tips; inflor [e]; single fl. [d]	almost hairless; in chalk grassland; stigmas 2 or 3	*Phyteuma orbiculare** Round-headed Rampion
	broad bracts below fl. head [f] single fl. [c]	downy/hairy; not on chalk	*Jasione montana* Sheep's-bit
Small creeping plant; lvs. roundish, lobed		in damp, acid places	*Wahlenbergia hederacea** Ivy-leaved Bellflower
Stem lvs. mostly linear	fl. [a]		*Campanula rotundifolia* Harebell
Fls. almost stalkless, in clusters	sepals longer than petals [g]		*Legousia hybrida* Venus's-looking-glass
	corolla much larger than calyx [b]	calcareous soil	*Campanula glomerata* Clustered Bellflower

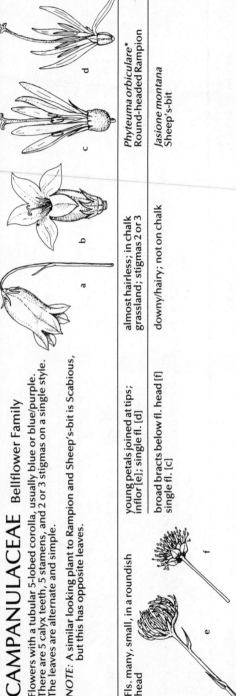

173

▶

Calyx tube hairy	sepals erect; stem sharply angled	50–100 cm high	*Campanula trachelium* Nettle-leaved Bellflower
	sepals spreading; stem hardly angled	30–60 cm high; stem lvs. stalkless	*Campanula rapunculoides* Creeping Bellflower
Calyx tube almost hairless	corolla 40–55 mm	50–120 cm high; lower lvs. w. winged stalks	*Campanula latifolia* Giant Bellflower

LOBELIACEAE Water Lobelia

The flowers have a tubular corolla with 5 irregular lobes.
An aquatic plant with a basal tuft of linear leaves, and a few bracts on the flower stem.

| Aquatic w. linear lvs. | fls. pale lilac | in acid mt. lakes | *Lobelia dortmanna* Water Lobelia |

RUBIACEAE Bedstraw Family

A distinctive family whose members have square stems and leaves in whorls of 4 to 12.
The flowers are small, having a 4- or 5-lobed corolla and a calyx which is sometimes no more than a low ridge.

Fls. lilac/pink	lvs. finely toothed [a]; sepals distinct		*Sherardia arvensis* Field Madder
	lvs. linear [b]; petals v. pale	calcareous grassland & dunes	*Asperula cynanchica* Squinancywort
Fls. yellow to yellowish-green (All the following fls. are whitish)	petals & stamens 4	lvs. linear, 6–12 in a whorl	*Galium verum* Lady's Bedstraw
		lvs. hairy, 4 in a whorl	*Galium cruciata* Crosswort

Lvs. w. 3 veins	petals & stamens 5		lvs. edged w. prickles; S. & W. only	*Rubia peregrina* Wild Madder
		northern plant		*Galium boreale* Northern Bedstraw
Corolla tube at least as long as corolla lobes	lvs. linear, 4–6 in a whorl [a]		in calcareous grassland & dunes	*Asperula cynanchica* Squinancywort
	lvs. lanceolate, 6–8 in a whorl		in woods	*Galium odoratum* Woodruff
Stem angles v. rough w. tiny prickles	lvs. w'out a terminal bristle		lvs. [b] 4–5 in a whorl; in wet places	*Galium palustre* Common Marsh-bedstraw
			lvs. 5–7 in a whorl; in dry places	*Galium parisiense** Wall Bedstraw
	fls. 1–3 in a cluster; fr. stalks bent down		cornfields, etc. usu. calcareous	*Galium tricornutum** Corn Cleavers
	marsh plant		lvs. [c] 6–8 in a whorl	*Galium uliginosum* Fed Bedstraw
	fr. covered w. hooked bristles; lvs. 12–50 mm		hedges, waste places	*Galium aparine* Cleavers
	fr. nearly smooth; lvs. 5–10 mm, often reflexed		walls, sandy places	*Galium parisiense** Wall Bedstraw
Teeth on lf. margin point backwards	lvs. usu. 4–5 in a whorl		wet places	*Galium palustre* Common Marsh-bedstraw
	lvs. usu. 6–8 in a whorl		grassy, often calcareous places	*Galium pumilum* Slender Bedstraw
low, mat-forming plant; lvs. 7–10 mm			heathlands	*Galium saxatile* Heath Bedstraw
woodland plant, 15–45 cm high	lvs. 25–40 mm			*Galium odoratum* Woodruff

| Robust hedgerow plant up to 120 cm | lvs. 8–25 mm | | *Galium mollugo* Hedge Bedstraw |

CAPRIFOLIACEAE Honeysuckle Family

Except for Twinflower and Dwarf Elder these are all obviously shrubs (or even small trees).
The clusters of flowers are followed by succulent fruit (not Twinflower).
The corolla is 5-lobed and there are 5 stamens.
Leaves are in pairs.

Lvs. pinnate	shrub or small tree	inflor. flat-topped; ripe fr. black	*Sambucus nigra* Elder
		inflor. ovoid; ripe fr. red	*Sambucus racemosa** Red-berried Elder
	stems not woody; 60–120 cm high	fr. black	*Sambucus ebulus* Dwarf Elder
Low creeping plant up to 15 cm high	fls. pink, in pairs, long-stalked	Scotland, mostly in pinewoods	*Linnaea borealis** Twinflower
Fls. tubular, 4–5 cm long	twining or scrambling shrub	fr. red	*Lonicera periclymenum* Honeysuckle
Lvs. almost entire	fls. pink; fr. white		*Symphoricarpos rivularis* Snowberry
Fls. in large umbel-like clusters	outer fls. much larger than others; fr. red	lvs. lobed and toothed [b]	*Viburnum opulus* Guelder-rose
	fls. all alike; ripe fr. black	lvs. ovate, toothed [a]; calcareous soil	*Viburnum lantana* Wayfaring-tree

ADOXACEAE Moschatel

An inconspicuous little plant with an unusual arrangement of 5 flowers on the end of a stem. The top flower has 4 petals and the others 5.
The radical leaves are well lobed, often into 3's and then 3's again.

Fls. greenish, 5 in a head	up to 10 cm high	woods and mt. ledges

Adoxa moschatellina
Moschatel

VALERIANACEAE Valerian Family

Plants with clusters of small flowers, pinkish in *Valeriana*, pale lilac-blue in *Valerianella*.
The funnel-shaped flowers have a 5-lobed corolla and 3 stamens (only 1 stamen in Red Valerian, and none in female flowers of Marsh Valerian).
The calyx is no more than a rim beneath the corolla.
NOTE: Ripe fruit are needed to name the species of *Valerianella*.

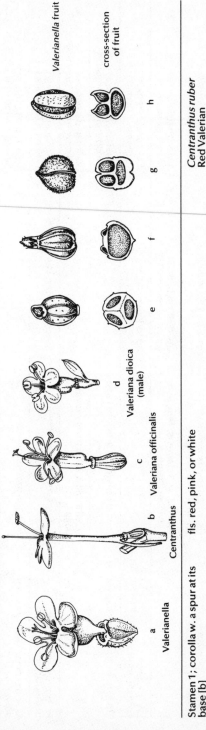

a
Valerianella

b
Centranthus

c
Valeriana officinalis

d
Valeriana dioica
(male)

e

f

g

h

Valerianella fruit

cross-section
of fruit

Stamen 1; corolla w. a spur at its base [b]	fls. red, pink, or white	*Centranthus ruber*
Red Valerian

Some lvs. pinnate	fls. w. 3 stamens and a style [c] on previous page	usu. 30–120 cm high	*Valeriana officinalis* Common Valerian
Lvs. ovate or cordate, toothed	male fls. w. 3 stamens; female w. 1 style, no stamens	up to 30 cm high [d] on previous page	*Valeriana dioica* Marsh Valerian
		up to 150 cm high	*Valeriana pyrenaica** Pyrenean Valerian
Tip of fr. bears minute cup (calyx)	fr. w. 3 cells [e] on previous page		*Valerianella rimosa** Broad-fruited Cornsalad
	fr. w. 1 cell & 2 ribs [f] on previous page		*Valerianella dentata* Narrow-fruited Cornsalad
Fr. approx. spherical to triangular, compressed	fr. [g]; fl. [a] both on previous page	the commonest *Valerianella* [a]	*Valerianella locusta* Common Cornsalad
Fr. longer than wide, grooved down one side	fr. [h] on previous page		*Valerianella carinata** Keeled-fruited Cornsalad

DIPSACACEAE Teasel and Scabious

Bluish or white flowers arranged in compact heads.
Each floret, which may be 4- or 5-lobed, has 4 protruding stamens.
The leaves are opposite.

There are similar-looking flowers in the Campanulaceae (p. 173) but these have alternate leaves and stamens that are inside the corolla.

Stems stout, prickly; fl. heads w. spiny bracts	fls. purple; fl. head 3–8 cm long	fl. head always erect [a]	*Dipsacus fullonum* Teasel
	fls. white; head spherical about 2 cm diameter	fl. head at first drooping [b]	*Dipsacus pilosus* Small Teasel
Corolla w. 4 lobes	stem lvs. deeply lobed	fl. head bracts broad, in 2 rows [c]	*Knautia arvensis* Field Scabious

	stem lvs. hardly lobed		*Succisa pratensis* Devil's-bit Scabious
	fl. head bracts narrow [d] on previous page; lvs. w. narrow lobes		*Scabiosa columbaria* Small Scabious
Corolla 5-lobed	damp places	dry calcareous places	

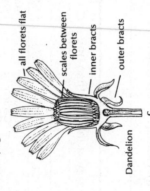
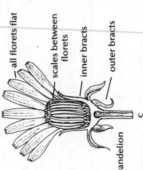

tubular floret d

floret with flat lobe e

feathery pappus g

pappus — beak f — fruits

COMPOSITAE Daisy Family

A very large family whose tiny flowers (florets) are typically massed into compact heads. Each floret may be tubular [d] as in the central disc of a Daisy, or may bear a flat lobe [e] as in the Daisy's outer white rays. In some species both types of floret are present (e.g. Daisy), in others there may be only the flat-lobed type (e.g. Dandelion), or only the tubular ones (e.g. Thistles). In a few species the florets may be small in number or loosely arranged.

The 5 stamens are all joined together to form a sleeve round the style and are not very obvious.

The style is often longer than the corolla tube and has 2 stigmas.

Beneath each head of florets there are often sepal-like bracts.

The fruit is one-seeded and may bear a pappus [f and g].

Teasels and Scabious have 4 separate and protruding stamens to each floret (page 178)

Sheep's-bit and Rampion (page 173).

ray floret

disc floret

space between florets

receptacle

Daisy type a

ray floret

disc floret

inner bracts

outer bracts

Ragwort b

all florets flat

scales between florets

inner bracts

outer bracts

Dandelion c

Each fl. head like a daisy, w. yellow centre and white rays [a]   Section 1 page 181

Each fl. head w. compact daisy-like centre and yellow rays [b]   Section 2 page 182

179

Fls. entirely yellow/orange; all florets flat like those of dandelion [c] on previous page / See also the Simplified Key on page 260	Section 3 page 185
Lvs. prickly, like a thistle; fls. usually purple	Section 4 page 189
Fls. in small greenish, green/brown, green/yellow, or yellowish clusters; no ray florets; fl. heads up to about 6 mm diameter	Section 5 page 190
Other plants	Section 6 page 192

COMPOSITAE List of genera (Some genera may appear in more than one section)

Section	1	2	3	4	5	6
Flowers like	Daisy	Ragwort	Dandelion	Thistle	Groundsel	Others
	Achillea	Chrysanthemum	Arnoseris	Carduus	Artemisia	Achillea
	Antennaria	Doronicum	Crepis	Carlina	Filago	Anaphalis
	Anthemis	Inula	Hieracium	Centaurea	Gnaphalium	Antennaria
	Bellis	Pulicaria	Hypochaeris	Cirsium	Matricaria	Arctium
	Chamaemelum	Senecio	Lactuca	Onopordum	Senecio	Aster
	Chrysanthemum	Solidago	Lapsana	Silybum	Tanacetum	Bidens
	Conyza	Tussilago	Leontodon		Xanthium	Centaurea
	Erigeron		Mycelis			Cichorium
	Galinsoga		Picris			Cirsium
	Leucanthemum		Sonchus			Erigeron
	Matricaria		Taraxacum			Eupatorium
	Tripleurospermum		Tragopogon			Inula
			Tussilago			Matricaria
						Petasites
						Saussurea
						Senecio
						Serratula
						Tanacetum
						Tragopogon
Page	181	182	185 (simplified key on p. 260)	189	190	192

COMPOSITAE Section 1 Flowers like a daisy with yellow centre and white rays

All lvs. radical

Lvs. simple, entire or toothed, but not greatly divided	fl. heads solitary on long stalks			*Bellis perennis* Daisy
		fl. head 3–6 cm across		*Leucanthemum vulgare* Oxeye Daisy
			fl. head size of Daisy; naturalized in S. & S.W.	*Erigeron mucronatus** Mexican Fleabane
	lvs. opposite; only 4–5 white rays [a]		scales between florets 3-pronged [e]	*Galinsoga parviflora* Gallant Soldier
			scales between florets simple [f]; stem v. hairy [a]	*Galinsoga ciliata* Shaggy Soldier
	stem not branched; lvs. narrow, woolly beneath; inflor. terminal		plant 30–100 cm tall [b]; lvs. 6–10 cm	*Anaphalis margaritacea* Pearly Everlasting
			up to 20 cm tall; lvs. 1–4 cm; hills & mts [b]	*Antennaria dioica* Mountain Everlasting
	lvs. clearly toothed; fl. head 12–18 mm across		[c]	*Achillea ptarmica* Sneezewort
	fl. head about 5 mm diam. ray florets almost erect		[d]	*Erigeron canadensis* Canadian Fleabane
Fl. heads up to 6 mm across, in close clusters	lvs. finely divided, dark green, aromatic		fls. occasionally pinkish	*Achillea millefolium* Yarrow

Some scales between disc florets	lvs. hardly scented, downy beneath; scale as [a]	*Anthemis arvensis* Corn Chamomile
	foetid-smelling; disc scales linear [b]	*Anthemis cotula* Stinking Chamomile
	strongly & pleasantly scented; scales broad [c]	*Chamaemelum nobile* Chamomile
Lvs. v. finely divided	plant aromatic	
	receptacle hollow compare [f]	*Matricaria recutita* Scented Mayweed
	receptacle solid	
	fr. w. 2 round glands [d]; widespread	*Tripleurospermum inodorum* Scentless Mayweed
	fr. w. long ovate glands [e]; lvs. fleshy; maritime	*Tripleurospermum maritimum* Sea Mayweed
Fl. heads few, 3–6 cm across		*Leucanthemum vulgare* Oxeye Daisy
Plant yellowish-green, aromatic	fl. heads 12–20 mm in flat topped clusters	*Tanacetum parthenium* Feverfew

COMPOSITAE Section 2 Each flower head like a daisy but entirely yellow

Stems w. many scales but no true lvs.	fls. appear before lvs. Mar.–Apr.	*Tussilago farfara* Colt's-foot
Shrub w. grey downy deeply lobed lvs.	S. & W. sea cliffs	*Senecio bicolor* Silver Ragwort
Lvs. succulent, linear often 3-toothed at tip	cliffs & saltmarshes	*Inula crithmoides* Golden Samphire

disc floret

a b c

d e mature fruit

receptacle f

Lvs. w. cordate base and/or clasping stem	fl. heads 4–6 cm across; upper lvs. ovate; clasping stem	*Doronicum pardalianches* Leopard's-bane
	fl. heads 1.5–3 cm; lvs oblong, wrinkled clasping stem	*Pulicaria dysenterica* Common Fleabane
Fl. head 3.5 cm or more across	plant 60–150 cm tall; lvs. large, ovate, toothed	*Inula helenium* Elecampane
	lvs. bluish, green, lobed, rather fleshy	*Chrysanthemum segetum* Corn Marigold
Ray florets none or few or under 5 mm	lvs. lanceolate, w. only small teeth; 30–130 cm tall	dry calcareous places lf. and fl. head [a] — *Inula conyza* Ploughman's-spikenard
	no ray florets at all [b]	*Senecio vulgaris* Groundsel
	fl. head bracts w. black tips, outer ones v. short	greyish-yellow/green plant usu. over 30 cm fl. head [c] — *Senecio sylvaticus* Heath Groundsel
		smaller plant — *Senecio vulgaris* var. *radiata* Rayed Groundsel
	outer fl. head bracts nearly ½ length of inner [d]	v. sticky, glandular and foetid — *Senecio viscosus* Sticky Groundsel
Lvs. mostly divided into rather narrow segments; fl. heads 15–30 mm across	marsh plant w. large lobe at lf. end [e]	fl. heads 2.5–3 cm across; ray florets 3-toothed — *Senecio aquaticus* Marsh Ragwort
	ray florets 3-toothed; fl. hds. 15–25 mm across	outer fl. hd. bracts about ½ as long as inner; lf. [f] — *Senecio erucifolius* Hoary Ragwort
		outer fl. hd. bracts less than ½ as long as inner — *Senecio jacobaea* Common Ragwort
	ray florets notched or nearly entire	lvs. hairless, variously lobed — *Senecio squalidus* Oxford Ragwort

Fl. head bracts in one row, all equal [a]		in calcareous grassland	Senecio integrifolius Field Fleawort
Fl. head bracts in 2 rows, outer smaller than inner	lvs. long & pointed w. small teeth [b]	in wet places	Senecio fluviatilis* Broad-leaved Ragwort
	usu. some lvs. lobed [c]		Senecio squalidus Oxford Ragwort
Ray florets spreading [d]			Solidago virgaurea Goldenrod
Ray florets short, almost erect [e]		in wet places	Pulicaria vulgaris* Small Fleabane

fl. heads

leaves

a	b	c	d	e
Senecio integrifoliu	Senecio fluviatilis	Senecio squalidus	Solidago virgaurea	Pulicaria vulgaris

COMPOSITAE Section 3 Flower heads entirely yellow/orange. All florets flat like a Dandelion.

Hawkweeds (*Hieracium*) and Dandelions (*Taraxacum*) produce seed without the need for pollination, and so give rise to a very large number of microspecies, which cannot be considered here.

See also the Simplified Key and drawings on page 260.

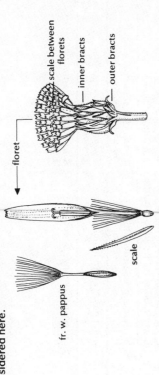

Stems unbranched w. many scales but no true lvs.	fls. appear before lvs.	*Tussilago farfara* Colt's-foot
Fls. orange/brown		*Hieracium aurantiacum* Fox-and-Cubs
All lvs. grasslike	fl. hd. bracts longer than florets	*Tragopogon pratensis* Goat's-beard
Lvs. usu. w. obvious spots	lvs. pimply & bristly	*Picris echioides* Bristly Oxtongue
	pappus feathery; scales between florets [a]	*Hypochaeris maculata** Spotted Car's-ear
	pappus simple; no scales among florets [b]	(includes many micro-species) *Hieracium murorum* (agg.) Hawkweed

▶

Fl. head solitary on unbranched stem			
stem hollow throughout w. milky juice	lvs. simple w. long hairs [a]		*Hieracium pilosella* Mouse-ear Hawkweed
	lvs. mostly toothed (includes many micro-species)		*Taraxacum officinale* (agg.) Dandelion
stem swollen & hollow just below fl. head	fl. head 7–11 mm across; no pappus on fr. fl. hd. [b]		*Arnoseris minima** Lamb's Succory
	fl. hd. 25–40 mm across; fr. w. pappus [c, d]		*Leontodon hispidus* Rough Hawkbit
lvs. whitish beneath w. long hairs [a]			*Hieracium pilosella* Mouse-ear Hawkweed
small bracts on stem	stem stout, rough, hairy; fl. hd. 25–40 mm across [c, d]		*Leontodon hispidus* Rough Hawkbit
	scales between florets [e]; lvs. hairy		*Hypochaeris radicata* Cat's-ear (stunted)
	scales between florets; inner & outer fr. differ [f]; lvs almost hairless		*Hypochaeris glabra* Smooth Cat's-ear
	lvs. w. deep & narrow lobes fr. [g]		*Leontodon autumnalis* Autumn Hawkbit
fl. hd. 12–20 mm; outer florets w' out pappus [h]	outer petals grey/violet beneath		*Leontodon taraxacoides* Lesser Hawkbit
fl. hd. 25–40 mm across	outer petals orange beneath [c]; fr [d]		*Leontodon hispidus* Rough Hawkbit

a, b, c, d, e (inner / outer), f, g, h (inner / outer)

Stems v. bristly	fl. hd. w. 3–5 broad outer bracts [a]	lvs. w. pimples; fr. w. long beak	*Picris echioides* Bristly Oxtongue
	outer fl. hd. bracts narrow, spreading [b]	fr. not beaked	*Picris hieracioides* Hawkweed Oxtongue
Stem lfless but bearing tiny bracts	narrow scales between florets; fr. w. pappus [c]	lvs. usu. hairy	*Hypochaeris radicata* Cat's-ear
		lvs. almost hairless; petals hardly longer than bracts	*Hypochaeris glabra* Smooth Cat's-ear
	no pappus on fr.	stem swollen in upper part [e]	*Arnoseris minima** Lamb's Succory
	lvs. usu. deeply & narrowly lobed	fr. w. pappus [d]	*Leontodon autumnalis* Autumn Hawkbit
No pappus on fr.	usu. tall, slender, well branched		*Lapsana communis* Nipplewort
Usu. only 5 ray florets per head	stem w. milky juice; fr. hardly beaked		*Mycelis muralis* Wall Lettuce
Stems w. copious milky juice	fr. w. long beak	upper lvs. linear, w'out spines [g]	*Lactuca saligna** Least Lettuce
		lvs. ovate, margins entire w'out spines	*Lactuca sativa* Garden Lettuce
		stem bracts w. pointed auricles [h]; fr. ribbed [f]	*Lactuca serriola* Prickly Lettuce
		stem bracts rounded, appressed [i]; fr. almost smooth	*Lactuca virosa* Great Lettuce

▶

fl. hd. 4–5 cm diameter	buds hairy, w. sticky yellow glands [a]; lf. base [b]	*Sonchus arvensis* Perennial Sow-thistle
stem lvs. w. rounded but toothed auricles [c]	lvs. shiny	*Sonchus asper* Prickly Sow-thistle
stem lvs. w. pointed auricles [d]	lvs. dark green, dull	*Sonchus oleraceus* Smooth Sow-thistle
Pappus pale brownish (not white)		
fl. hd. bracts in 1 row w. a few small outer ones [e]	fls. orange/yellow	*Crepis paludosa* Marsh Hawk's-beard
fl. hd. bracts in 2 or more rows [f]	(includes many micro-species)	*Hieracium murorum* (agg.) Hawkweed
Pappus white		
fr. w. long beak [g]; stem purplish below	outer fl. hd. bracts spreading; fls. May–July	*Crepis vesicaria* Beaked Hawk's-beard
lvs. hardly toothed [i]	northern hills	*Crepis mollis** Northern Hawk's-beard
inner fl. hd. bracts downy inside	robust, hairy plant	*Crepis biennis* Rough Hawk's-beard
lower lvs. w. well-spaced teeth; upper lvs. linear; fr. [h]	usu. a slender plant; common; fls. June–Sep.	*Crepis capillaris* Smooth Hawk's-beard
most lvs. lanceolate toothed, w. long auricles	wet places in N. & Midlands	*Crepis paludosa* Marsh Hawk's-beard

COMPOSITAE Section 4 Thistle-like plants, normally with purple flowers

Lvs. w. conspicuous white veins	40–120 cm high		*Silybum marianum* Milk Thistle
Ring of straw-coloured bracts below fls.	10–60 cm high	calcareous grassland & dunes	*Carlina vulgaris* Carline Thistle
Fl. heads usu. solitary or well separated	fl. hd. at ground level		*Cirsium acaule* Dwarf Thistle
	stem lvs. deeply lobed and spiny	fl. hd. 3–5 cm diameter, often drooping	*Carduus nutans* Musk Thistle
	lvs. glabrous above, downy beneath; stem grooved	hardly spiny at all; N. of the Thames only [a]	*Cirsium helenioides* Melancholy Thistle
	lvs. hairy, green above, white beneath [b]	plant not very spiny	*Cirsium dissectum* Meadow Thistle
Fl. heads drooping; 3–5 cm diameter			*Carduus nutans* Musk Thistle

The species following are normally branched but poor specimens may have only one head of flowers.

Fl. hd. w. spreading spines, 2–2.5 cm long	all lvs. narrow or w. narrow lobes	upper lvs. usu. have a few teeth; fls. purplish pink	*Centaurea calcitrapa** Red Star-thistle
Lvs. woolly on both sides	stem w. continuous spiny wings		*Onopordum acanthium* Cotton Thistle
Pappus hairs feathery (each hair is again long-haired)	lvs. hairy/prickly above	fl. hd. v. large, woolly beneath [e]; calcaerous soil	*Cirsium eriophorum* Woolly Thistle
		fl. hds. 3–5 cm diameter [c] up to 3 in a cluster	*Cirsium vulgare* Spear Thistle
[Examine with a lens when dry]		fl. hds. smaller [d] in more crowded clusters	*Cirsium palustre* Marsh Thistle

▶

189

Fl. hds. cylindrical w. lanceolate bracts	fls. pale purple, fragrant as [d] on previous page	*Cirsium arvense* Creeping Thistle
	usu. pale purple or pinkish	*Carduus tenuiflorus* Slender Thistle
Fl. hds. oval/spherical w. linear bracts	fls. dark purple	*Carduus acanthoides* Welted Thistle

COMPOSITAE Section 5 Flowers in small green/brown or green/yellow clusters without ray florets

Flower heads up to about 6 mm in diameter.
For similar flowers with larger heads or a loose inflorescence see Section 6 p. 192. Some species appear in both sections.

Plant w. long 3-forked spines	hds. in fr. about 10 mm	*Xanthium spinosum* Spiny Cocklebur
Fl. hds. yellow w. a few small rays	lvs. grey-green	*Senecio sylvaticus* Heath Groundsel
Lvs. many-lobed, mostly aromatic	fl. hd. conical, hollow; like a Mayweed w' out rays [a]	
	5–30 cm high; lvs. pineapple-scented lf. at [a]	*Matricaria matricarioides* Pineappleweed
	usu. under 45 cm high; fl. hd. cylindrical, yellow [b]	
	lf. at [b]	*Senecio vulgaris* Groundsel
	lvs. dark green above, whitish below; 60–120 cm high	
	in fl. July–Sep. inflor. & lf. at [c]	*Artemisia vulgaris* Mugwort
	in fl. Oct.–Nov., S.E. England	*Artemisia verlotiorum* Chinese Mugwort

a	lvs. silky both sides; aromatic; fls. yellowish	end lf. segments 2–3 mm wide	*Artemisia absinthium* Wormwood
		end lf. segments about 1 mm wide; maritime	*Artemisia maritima* Sea Wormwood
	lvs. pinnate, sharply toothed [a]	aromatic	*Tanacetum vulgare* Tansy
Terminal fl. clusters overtopped by lvs.	terminal clusters w. only 1–2 overtopping lvs.	fl. bracts red-tipped	*Filago lutescens** Red-tipped Cudweed
b	fl. hds. within each cluster flat-topped [b]	common [b]	*Gnaphalium uliginosum* Marsh Cudweed
	fl. hds. within each cluster ovoid		*Filago pyramidata** Broad-leaved Cudweed
Stems unbranched	plant up to about 12 cm	Scottish mts. only	*Gnaphalium supinum* Dwarf Cudweed
	plant erect, 8–60 cm		*Gnaphalium sylvaticum* Heath Cudweed
Fl. hds. in clusters near ends of branches	c	fl. hds. up to 6 in a cluster	*Filago minima* Small Cudweed
		fl. hds. usu. 10 or more in a cluster [c]	*Filago vulgaris* Common Cudweed

COMPOSITAE

Section 6 Various types of flower head, including many purplish ones without prickly leaves, yellowish ones without ray florets, and a few others

Fls. bright blue	fl. hds. terminal	lvs. linear/lanceolate, lower ones w. narrow lobes		*Centaurea cyanus** Cornflower
		lower lvs. w. broad terminal lobe; fls. lilac/blue		*Cicerbita macrophylla** Blue Sow-thistle
	fl. hds. at intervals up the stem			*Cichorium intybus* Chicory
Fls. creamy or white	lvs. finely divided, aromatic	inflor. at first sight like an umbel; occasionally pink		*Achillea millefolium* Yarrow
	lvs. linear/lanceloate toothed	ray florets broad, toothed [a]		*Achillea ptarmica* Sneezewort
		inflor. long, loose [b]		*Erigeron canadensis* Canadian Fleabane
		small plant w. rosette of lvs.	fl. hds. in an umbel [c]	*Antennaria dioica* Mountain Everlasting
			in fl. Dec.–March, almond-scented [d]	*Petasites fragrans* Winter Heliotrope
			bracts on fl. stem inrolled	*Petasites albus* White Butterbur
		stem bracts large, lf.-like, overlapping	garden escape	*Petasites japonicus** Giant Butterbur
	Fl. stem stout, hollow, w. lanceolate scales	inflor. a mass of small clusters	fls. usu. appear before lvs., in Spring	*Petasites hybridus* Butterbur
	In fl. Dec.–March, almond-scented [d]			*Petasites fragrans* Winter Heliotrope

Lvs. sheathing, grass-like; fls. purple		*Tragopogon porrifolius** Salsify
Fl. hds. yellowish		
lvs. simple, entire, woolly beneath, 6–10 cm	inflor. terminal, 6–10 cm high	*Anaphalis margaritacea* Pearly Everlasting
lvs. finely divided, green pineapple-scented [a]	5–30 cm high	*Matricaria matricarioides* Pineappleweed
lvs. pinnate, sharply toothed, v. aromatic [b]	30–100 cm high	*Tanacetum vulgare* Tansy
lower lvs. 3-lobed & toothed [c]	in wet places	*Bidens tripartita* Trifid Bur-marigold
lvs. lanceolate to linear-lanceolate, toothed	inflor. in spikes of tiny yellow clusters	*Solidago canadensis* Garden Goldenrod
	fl. hds. 15–25 mm across; in wet places	*Bidens cernua* Nodding Bur-marigold
	fl. hds. up to 12 mm across [d]; in dry places	*Inula conyza* Ploughman's-spikenard
lvs. succ. linear-lanceolate hairless [e]	in saltmarshes	*Aster tripolium* (rayless form) Sea Aster
fl. hds. 4–5 mm diameter [f]		*Senecio vulgaris* Groundsel
Stem lvs. entire, linear-lanceolate		
fls. purplish [g]	in saltmarshes lf. [e]	*Aster tripolium* Sea Aster
small plant w. lvs. woolly beneath	fls. white/pink; fl. hds. in an umbel [i]	*Antennaria dioica* Mountain Everlasting
lvs. hairy up to 7.5 cm; ray fl. pale purple	[h]	*Erigeron acer* Blue Fleabane

▶

Fl. hds. purple, surrounded by stiff hooks	fl. hd. stalks usu. over 4 cm; lf. stalks solid	fr. hd. usu. over 35 mm across	*Arctium lappa* Greater Burdock
	fl. hd. stalks usu. under 4 cm; lf. stalks hollow	fr. hd. 15–35 mm across	*Arctium minus* Lesser Burdock
Fl. hds. grouped in large clusters [d, e] on next page	fls. pale to dark reddish mauve	usu. 50–120 cm high, in moist places	*Eupatorium cannabium* Hemp-agrimony
Lvs. pinnate, sharply saw-toothed		[f] on next page	*Serratula tinctoria* Saw-wort
Lvs. lanceolate, woolly beneath	fl. hds. usu. solitary	N. of the Thames only [a]	*Cirsium helenioides* Melancholy Thistle
	fl. hds. in a small cluster	mts. in the N.	*Saussurea alpina* Alpine Saw-wort
Petals rise from a hard globular knob	upper lvs. lobed		*Centaurea scabiosa* Greater Knapweed
	fl. hd. bracts edged w. long fine teeth [b]		*Centaurea nigra* Common Knapweed
	fl. hd. bracts w. short blunt teeth [c]		*Centaurea jacea** Brown Knapweed
Fl. hds. lilac/blue, about 3 cm diameter	stem w. milky juice; lvs. w. only 1 pair of lobes		*Cicerbita macrophylla** Blue Sow-thistle

a

b   c

fl. hd. bracts

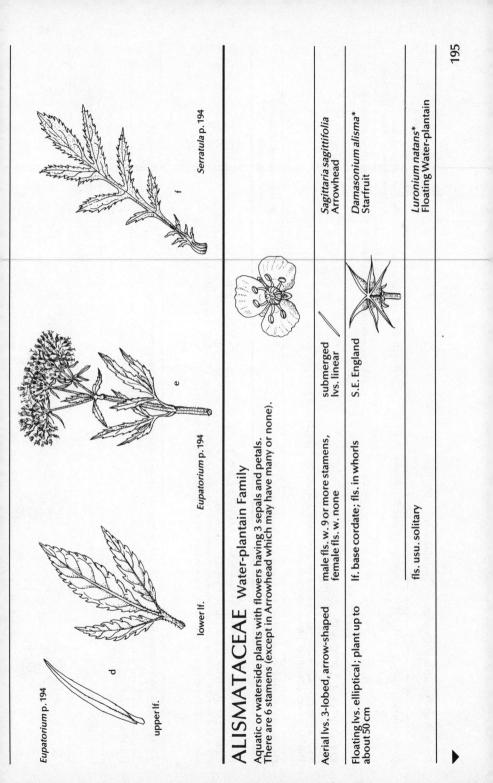

Eupatorium p. 194

upper lf.

lower lf.

Eupatorium p. 194

Serratula p. 194

ALISMATACEAE Water-plantain Family

Aquatic or waterside plants with flowers having 3 sepals and petals.
There are 6 stamens (except in Arrowhead which may have many or none).

Aerial lvs. 3-lobed, arrow-shaped	male fls. w. 9 or more stamens, female fls. w. none	*Sagittaria sagittifolia* Arrowhead
	submerged lvs. linear	
Floating lvs. elliptical; plant up to about 50 cm	If. base cordate; fls. in whorls	
	S.E. England	*Damasonium alisma** Starfruit
	fls. usu. solitary	*Luronium natans** Floating Water-plantain

Fls. in an umbel; up to 20 cm high	lvs. narrowly lanceolate		*Baldellia ranunculoides* Lesser Water-plantain
20–100 cm high	lf. tapers gradually to base	style inserted near top of fr. [a]	*Alisma lanceolatum* Narrow-leaved Water-plantain
	lf. base almost cordate	style inserted about ½-way down fr. [b]	*Alisma plantago-aquatica* Water-plantain

a   b

BUTOMACEAE Flowering-rush

A tall aquatic or waterside plant whose flowers have 3 sepals and petals and 6–9 stamens.
The leaves are all radical, linear, and 3-angled.

Fls. 25–30 mm, pink, in an umbel [c]	up to 150 cm high	*Butomus umbellatus* Flowering-rush

c

HYDROCHARITACEAE Frogbit Family

Aquatic plants whose flowers have 3 sepals and petals.
Male and female flowers are on separate plants, the females solitary, the males 1 to 3 together, with 9 to 12 stamens. Both sexes may also bear staminodes (like stamens without anthers).

Lvs. stiff, toothed, in a rosette [e]	fls. 3–4 cm across; petals white w. spot	in calcareous districts	*Stratiotes aloides* * Water-soldier
Lvs. in whorls of 3 [g]		rarely flowers	*Elodea canadensis* Canadian Waterweed
Lvs. broad, floating [f]	fls. about 2 cm across, white [d]		*Hydrocharis morsus-ranae* Frogbit

d   e   f   g

JUNCAGINACEAE Arrowgrasses

The small greenish flowers are in narrow spikes, the flower parts being arranged in 3's or 6's.
These are marsh plants with linear leaves.

The similar looking plantains have 4 long-stalked stamens to each flower (see page 172).

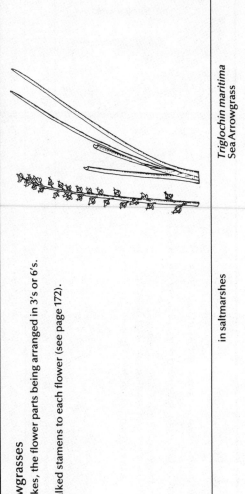

Fr. ovoid

in saltmarshes

Triglochin maritima
Sea Arrowgrass

Fr. cylindrical/tapering

in freshwater marshes

Triglochin palustris
Marsh Arrowgrass

ZOSTERACEAE Eelgrasses

The only British flowering plants that grow in the sea, below high-tide level.
The grass-like leaves are rooted in the sea bed in muddy places.
The flowers have either 1 stamen or 2 stigmas and are to be found in flat clusters fairly low down on the stems.

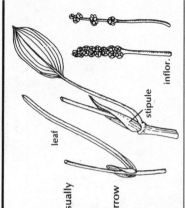

Lvs. 20–50 cm long, 5–10 mm wide [a]	fl. stem usu. branches	grows below low-tide level	*Zostera marina** Eelgrass
Lvs. 15–30 cm long, about 2 mm wide	older lvs. w. notched tip	exposed at low tide	*Zostera angustifolia** Narrow-leaved Eelgrass
Lvs. mostly under 12 cm long under 1 mm wide	older lvs. w. notched tip	exposed at low tide	*Zostera noltii** Dwarf Eelgrass

POTAMOGETONACEAE Pondweeds

Aquatic plants of fresh and brackish water.
The inflorescence is a spike of small greenish flowers, often rising above the surface.
The flower parts are all in 4's.
The leaves are very variable, depending greatly on the depth, speed and composition of the water, and usually have stipules at their base.
This variability, together with frequent hybridization, make this a difficult genus.
Because flowers and fruit are often absent this key relies almost entirely on leaf characters.

NOTE: Barren plants of the genera *Ruppia* (see page 201) and *Zannichellia* (page 202) are similar to the narrow leaved Potamogetons.

Lf. base widened into a sheath	inflor. becomes an umbel in fr.	only in coastal brackish pools	RUPPIACEAE p. 201
	If. sheath w. whitish edges [b]	lvs. dark green, mostly up to 2 mm wide	*Potamogeton pectinatus* Fennel Pondweed

Lvs. in pairs, lanceolate up to 25 mm long [a]

lvs. yellowish green, thread-like up to 1 mm wide | fr. whorls well spaced; in N. | *Potamogeton filiformis** Slender-leaved Pondweed

| *Groenlandia densa* Opposite-leaved Pondweed

Older lvs. frilly, toothed [b]

Potamogeton crispus Curled Pondweed

Floating lvs. ovate w. long stalk

submerged lvs. 1–3 mm wide floating lvs. [c] | *Potamogeton natans* Broadleaved Pondweed

submerged lvs. narrowly lanceolate

submerged lvs. net-veined [e]; only in S. | *Potamogeton nodosus** Loddon Pondweed

submerged lvs. stalkless, wavy [f] | *Potamogeton gramineus** Various-leaved Pondweed

all lvs. long-stalked, submerged ones narrower than floating [c] | *Potamogeton polygonifolius* Bog Pondweed

stalks often longer than lvs. [c] | in boggy places or acid water | *Potamogeton polygonifolius* Bog Pondweed

stalks shorter than lvs [d] | in calcareous water | *Potamogeton coloratus* Fen Pondweed

▶

Lvs. broad, blunt, clasping the stem [e]

Lvs. 2–6 cm long, oval [e]

Potamogeton perfoliatus
Perfoliate Pondweed

Lvs. 10 mm or more wide

stipules 2-keeled [d]

lf. stalks. v. short; no floating lvs. [c]

Potamogeton lucens
Shining Pondweed

lvs. reddish w. tapered base [a]

Potamogeton alpinus
Red Pondweed

lvs. green, stalkless, w. rounded base [b]

no floating lvs.

Potamogeton praelongus
Long-stalked Pondweed

Lvs. 3–12 mm wide, wavy, sessile, w. 7–11 veins

sometimes w. wider ovate floating lvs.

*Potamogeton gramineus**
Various-leaved Pondweed

Lvs. linear, 2–4 mm wide, w. 5 main veins

stem flattened, keeled; stipules blunt [f]

*Potamogeton compressus**
Grass-wrack Pondweed

stipules acute [g]

Potamogeton friesii
Flat-stalked Pondweed

Stems much flattened

usu. in calcareous water

lf.

*Potamogeton acutifolius**
Sharp-leaved Pondweed

Lvs. thread-like, usu. 1-veined [h]

lvs. up to 1 mm wide and 4 cm long [h]

*Potamogeton trichoides**
Hairlike Pondweed

Lvs. 0.5–4 mm wide; usu. 1 or 3 veins

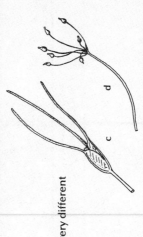

stipules 12–30 mm long, open to base [a]	lf. tip rounded w. tiny point	*Potamogeton obtusifolius* Blunt-leaved Pondweed
young stipules tubular at base [b]	midrib of older lvs. not bordered w. pale band	*Potamogeton pusillus* Lesser Pondweed
stipules 3–10 mm long, open to base	lvs. up to 2 mm wide	*Potamogeton berchtoldii* Small Pondweed

Lvs. 3-veined, under 0.5 mm wide   young stipules tubular at base [b]   *Potamogeton pusillus* Lesser Pondweed

RUPPIACEAE Tasselweed

Submerged aquatics of brackish coastal water.
The very slender leaves (about 0.5 mm wide) have a sheath at their base [c].
The minute flowers are in an umbel, the stalk of which lengthens considerably in fruit [d].
The slender-leaved Pondweeds (p. 201) and Horned Pondweed (p. 202) are similar in leaf but have very different inflorescences.

Fruiting umbel attached to a spirally twisted stalk 8 cm or more long	*Ruppia cirrhosa** Spiral Tasselweed
Fruiting umbel attached to a stalk up to c. 6 cm long; lf. apex acute	*Ruppia maritima* Beaked Tasselweed

201

ZANNICHELLIACEAE Horned Pondweed

A fine-leaved submerged aquatic, with stipules at the leaf bases.
The flowers are minute, in clusters in the leaf axils.
The slender-leaved Pondweeds (p. 201) and Tasselweed (p. 201) are similar in leaf but have very different inflorescences.

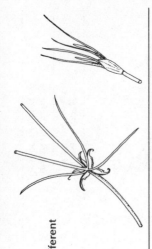

Lvs. usu. opposite, 0.5–2 mm wide: fr. like minute cucumbers in clusters of 2–6

Zannichellia palustris
Horned Pondweed

LILACEAE Lily Family

Flowers and inflorescences of varied form and colour.
The perianth of 6 segments may be petal- or sepal-like, with the segments free or united.
There are 6 stamens, except in the male flowers of Butcher's-broom, which have 3.
Some of the Alliums have tiny bulbs (bulbils) instead of or mixed with the flowers.

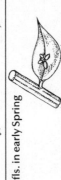

Dark spiny-lv'd shrub	fls. finy, either male or female; fr. a red berry	fls. in early Spring

Ruscus aculeatus
Butcher's-broom

Plant w. onion smell; fls. in a head, often mixed w. tiny bulbils	fls. white	lvs. linear keeled [a] inflor. [b]

Allium triquetrum*
Three-cornered Leek

		lvs. lanceolate [d] inflor. [c]

Allium ursinum
Ramsons

	common	

Allium vineale
Wild Onion

inflor. w. tiny bulbils but no fls. [e]

a	inflor. w. bracts much longer than the fl. hd.		*Allium oleraceum** Field Garlic
	stamens protrude; fls. usu. mixed w. bulbils [a]	common	*Allium vineale* Wild Onion
b	lvs. hollow; fls. purplish		*Allium schoenoprasum** Chives
	lvs. flat [b]; fls. purplish		*Allium scorodoprasum** Sand Leek
Well-branched; lvs. small, soft, linear	fl. greenish	fr. a red berry	*Asparagus officinalis* Wild Asparagus
Fls. yellowish to greenish-white; lvs. radical, linear	fls. long-stalked, in an umbel	only 1 radical lf. 15–40 cm long	*Gagea lutea** Yellow Star-of-Bethlehem
	stigma 3-lobed	radical lvs. in tuft, up to 4 cm [c]; hills in N.	*Tofieldia pusilla** Scottish Asphodel
c	stigma single; stamens hairy	in wet acid places	*Narthecium ossifragum* Bog Asphodel
Fls. 2.5 cm or more across, purplish	no lvs. at fl. time (Aug.–Oct.)		*Colchicum autumnale** Meadow Saffron
	petals bent back; stamens protrude		*Lilium martagon** Martagon Lily
	one fl. to a stem		*Fritillaria meleagris** Fritillary

▶

Fls. blue to purplish; lvs. linear		
fls. purplish, July to Sep.	mostly along S.W. coasts, in short grass	*Scilla autumnalis** Autumn Squill
fls. deep blue, egg-shaped, in dense head	mostly in E. Anglia and Oxon.	*Muscari neglectum** Grape Hyacinth
fl. stem 20–50 cm; fls. drooping	widespread	*Hyacinthoides non-scripta* Bluebell
under 15 cm high [a]	W. & N. coasts in fl. April–June	*Scilla verna* Spring Squill

Lvs. all or nearly all radical; fls. whitish		
fls. bell-like, drooping in one-sided spike	lvs. lanceolate; in woods	*Convallaria majalis* Lily-of-the-valley
fls. tiny w. 3-lobed bracts fls. & lvs. [b]	hills in the N.	*Tofieldia pusilla** Scottish Asphodel
petals green-striped, 15–20 mm long [c]	grassy places	*Ornithogalum umbellatum** Star-of-Bethlehem

Stem arching: fls. white, drooping

a

fls. 1–2 together; stem angular

*Polygonatum odoratum** Angular Solomon's-seal

fls. 2–5 together, constricted in middle [a]

Polygonatum multiflorum Solomon's-seal

TRILLIACEAE Herb-Paris

A plant with one flower on an upright stem.
There are 4 sepals and petals, both green, and 8 stamens.

Lvs. in a whorl of 4 just below fl.   fr. a black berry   in woods on calcareous soil

*Paris quadrifolia** Herb-Paris

JUNCACEAE Rushes and Woodrushes

The 3 sepals and 3 petals of Rush flowers are normally indistinguishable from each other and are all called perianth segments (per. seg.) see also page 6.

There are almost always 6 stamens and 3 stigmas to each flower.
The fruit is a capsule, containing either 3 or many seeds.
Leaves are linear and may be flat, channelled or tubular.

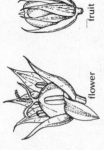

flower   fruit

Lvs. flat, grasslike, edged with fine hairs, never tubular; capsule contains 3 seeds	Section 1 (Luzula) p. 000
Lvs. tubular or channelled, round or elliptical in section, hairless; capsule contains many seeds	Section 2 (Juncus) p. 000

JUNCAEAE Section 1 Luzula Woodrushes

Fls. in ones (each on a stalk)	inflor. & fr. erect or one-sided	*Luzula forsteri* Southern Woodrush
	inflor. & fr. spreading and/or	*Luzula pilosa* Hairy Woodrush
Plant 30–80 cm high; lvs. over 6 mm wide		*Luzula sylvatica* Great Woodrush
Inflor. a short drooping spike	in N. & mts.	*Luzula spicata* Spiked Woodrush
Anthers at least twice as long as filaments [a]	up to about 15 cm high	*Luzula campestris* Field Woodrush
Anthers about equal to filaments [b]	20–40 cm high	*Luzula multiflora* Heath Woodrush

a   stamen

b   stamen

JUNCACEAE Section 2 Juncus (Rushes)

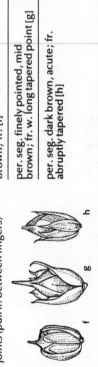

Lvs. & stems all alike; inflor. at side of stem	lvs. hollow w. many cross-joints	(pull lf. between fingers to feel joints) [f]	*Juncus subnodulosus* Blunt-flowered Rush
	stems v. fine, up to 30 cm; inflor. ½-way down	in N.	*Juncus filiformis** Thread Rush
	stiff maritime plant w. spine-tipped stems	up to 1 m; fr. hardly longer than per. seg. [a]	*Juncus maritimus* Sea Rush
		up to 1.5 m; ripe fr. much longer than per. seg. [b]	*Juncus acutus** Sharp Rush
	tufted plant	stem below inflor. smooth, glossy [c]	*Juncus effusus* Soft Rush
		inflor. compact; stem w. about 40 fine ridges [d]	*Juncus conglomeratus* Compact Rush
		inflor. loose; stem w. under 20 ridges [e]	*Juncus inflexus* Hard Rush
	creeping plant	in N.	*Juncus balticus** Baltic Rush
Lvs. hollow like a flattened tube	usu. over 40 cm high; lvs. w. cross-joints (pull lf. between fingers)	per. seg. obtuse, pale to mid brown; fr. [f]	*Juncus subnodulosus* Blunt-flowered Rush
		per. seg. finely pointed, mid brown; fr. w. long tapered point [g]	*Juncus acutiflorus* Sharp-flowered Rush
		per. seg. dark brown, acute; fr. abruptly tapered [h]	*Juncus articulatus* Jointed Rush

a b

c d e

f g h

▶

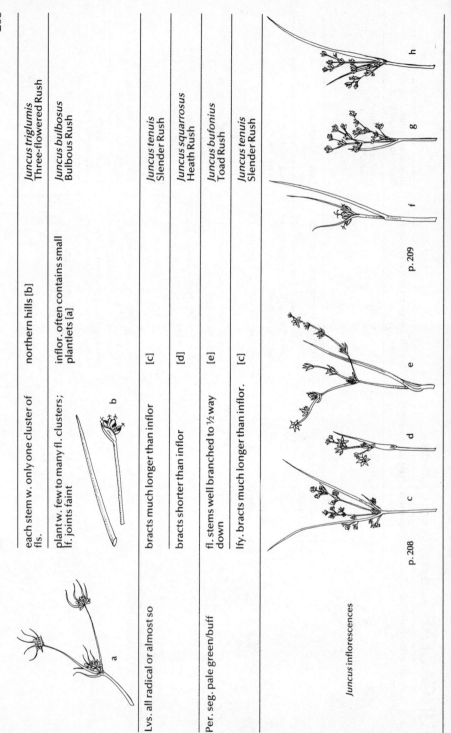

208

each stem w. only one cluster of fls.　　northern hills [b]　　*Juncus triglumis* Three-flowered Rush

plant w. few to many fl. clusters; lf. joints faint　　inflor. often contains small plantlets [a]　　*Juncus bulbosus* Bulbous Rush

Lvs. all radical or almost so

bracts much longer than inflor　　[c]　　*Juncus tenuis* Slender Rush

bracts shorter than inflor　　[d]　　*Juncus squarrosus* Heath Rush

Per. seg. pale green/buff

fl. stems well branched to ½ way down　　[e]　　*Juncus bufonius* Toad Rush

lfy. bracts much longer than inflor.　　[c]　　*Juncus tenuis* Slender Rush

a

b

c　　d　　e

p. 208

f　　g　　h

p. 209

Juncus inflorescences

Bracts v. much longer than compact inflor. [f] on previous page	10–30 cm high	*Juncus trifidus* Three-leaved Rush
Per. seg. dark brown; bracts shorter than inflor. [g] on previous page	saltmarshes	*Juncus gerardi* Saltmarsh Rush
Per. seg. light brown; bract longer than inflor. [h] on previous page	mostly in calcareous meadows	*Juncus compressus* Round-fruited Rush

AMARYLLIDACEAE Daffodil Family

The flower parts are in 6's or 2 sets of 3.
The flower buds are enclosed in a sheath.
The long, narrow leaves spring from the plant base.

Fls. yellow			*Narcissus pseudonarcissus* Wild Daffodil
Fls. white	only 1 fl. to a stem	fls. Feb.–April	*Galanthus nivalis* Snowdrop
	fls. 2 or more to a cluster	fls. April–May	*Leucojum aestivum** Summer Snowflake

IRIDACEAE Iris Family

Flowers with 6 perianth segments, sometimes arranged in 2 sets of 3.
There are 3 stamens, and 3 styles which may be large enough to resemble petals.
The leaves are long and narrow.

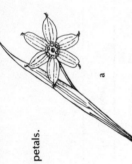

Petals orange, all similar	*Tritonia × crocosmiflora** Montbretia
Petals blue, all similar [a]	*Sisyrhinchium bermudiana** Blue-eyed grass
stamens fused to style	
Petals yellow, in 2 sets of 3	*Iris pseudacorus* Yellow Iris
styles (w. stigmas) resemble petals	
Petals purple in 2 sets of 3	*Iris foetidissima* Stinking Iris
styles (w. stigmas) resemble petals	

DIOSCOREACEAE Black Bryony

A climbing, twining plant with heart-shaped leaves.
Each flower has 6 perianth segments containing either 6 stamens or 3 stigmas.
Male and female flowers are on separate plants.

Fls. greenish yellow 4–5 mm across [b]	*Tamus communis* Black Bryony
fr. a berry, green to orange to red, poisonous	

ORCHIDACEAE Orchid Family

A family with a fascinating variety of flower form. The inflorescence is a simple spike, whose flowers have 3 sepals and 3 petals usually all coloured. The lower petal forms a lip, sometimes of extraordinary shape, and may also be elongated into a spur at the back.
The single anther is split into two pollen masses called pollinia, which may be carried away by an insect.
There is a small bract at the base of each flower.
Leaves are always simple and untoothed, with parallel veins.
NOTE: Because orchids do not produce flowers every year and because the time needed to produce the first flower spike from seed may be anything from 5 to 15 years *no whole flowering spike should ever be picked.*

Similar-looking flowers may be found in these families:
MONOTROPACEAE p. 148 but these flowers have 4–5 sepals and equal petals.
SCROPHULARIACEAE p. 159 but these flowers have 4 or 5 stalked stamens, and net-veined leaves.
OROBANCHACEAE p. 164 but the flowers have 4 stamens and a 2-lipped, 5-lobed corolla.
LABIATAE p. 166 but these have 4-angled stems and opposite leaves.

Plant without normal lvs., only yellowish brown scales on stem	lower corolla lip long, w. 2 spreading lobes		in shady woods (compare *Monotropa* p. 148)	*Neottia nidus-avis* Bird's-nest Orchid
	lower lip w. 1 large & 2 small lobes [a]		in N. only	*Corallorhiza trifida** Coralroot Orchid [a]
Usu. only 2 lvs. in an almost opposite pair; sometimes much smaller stem lvs. present	fls. whitish w. a long thin spur [b]		pollinia parallel [d], fls. 11–18 mm across	*Platanthera bifolia* Lesser Butterfly-orchid [b]
			pollinia not parallel [e]; fls. 18–23 mm across	*Platanthera chlorantha* Greater Butterfly-orchid
	lower corolla lip 2-lobed [c]; lvs. ovate		6–20 cm high, w. 6–12 red-tinged fls.	*Listera cordata** Lesser Twayblade
			20–60 cm high; fls. yellow/green, many [c]	*Listera ovata* Common Twayblade [c]

▶

211

lvs. about 1 cm long, usu. w. tiny buds at tip	usu. grows among *Sphagnum*; fls. yellow-green [a]		*Hammarbya paludosa** Bog Orchid [a]
	lower lip 3-lobed, yellow-green	calcareous grassland [b]	*Herminium monorchis** Musk Orchid [b]
	lip broad, often bent upwards, yellow-green	fens & dune slacks [c]	*Liparis loeselii** Fen Orchid [c]
Lvs. about 1 cm long, usu. w. tiny buds at tip	3–12 cm high, usually among *Sphagnum* [a]	fls. yellow/green	*Hammarbya paludosa** Bog Orchid [a]
Fls. white, arranged in a spiral; plant up to 25 cm	lvs. w. clear cross-veins; stem partly creeping	pinewoods in N. and Norfolk fl. [d]	*Goodyera repens* Creeping Lady's-tresses [d]
	stem upright; lower lip w. fringed margin [e]	grassy places, mostly in S.	*Spiranthes spiralis* Autumn Lady's-tresses [e]
Fls. greenish; mid lobe of lip 3–5 cm long [f]		mostly in S.E.	*Himantoglossum hircinum* Lizard Orchid [f]
Spur v. slender, about 12 mm long; fls. pinkish	spike dense, conical	calcareous grassland	*Anacamptis pyramidalis* Pyramidal Orchid [g]

a b c

d e f g

spike cylindrical; sweet-scented — *Gymnadenia conopsea* Fragrant Orchid [a]

Lower lip w. 4 narrow lobes, yellowish [b] — lobes like a man's arms & legs — *Aceras anthropophorum** Man Orchid [b]

Mid lobe of lower lip brown, downy — sepals green — lip about as wide as long w. central notch [c] — *Ophrys sphegodes** Early Spider-orchid [c]

lower lip distinctly lobed [e] — *Ophrys insectifera** Fly Orchid [e]

sepals pink [d] — *Ophrys apifera* Bee Orchid [d]

a — spur

b

c

d

e

Fls. 2–3 mm, greenish white, w. 2–3 mm spur [f] — lower lip 3-toothed; mostly in Wales & N. — *Pseudorchis albida** Small-white Orchid [f]

f — spur

▶

Spur about 2 mm; upper corolla lobes hooded	fls. green, tinged red/brown	lip w. 2 large & 1 small lobe [a]		*Coeloglossum viride* Frog Orchid [a]
		lip 4-lobed [b], white, spotted	sepals at first maroon	*Orchis ustulata** Burnt Orchid [b]
Fls. w. spur †	all perianth segments except lip form a hood	sepals green-veined; lips purplish [c]		*Orchis morio* Green-winged Orchid [c]
			lip pink w. 2 narrow & 2 broad lobes [d]; S.E.	*Orchis purpurea** Lady Orchid [d]
		spur at least as long as ovary, often curved up [e]		*Orchis mascula* Early-porple Orchid [e]
		lvs. usu. spotted; stem solid or almost so	lip w. 3 well separated lobes [h]	*Dactylorhiza fuchsii* Common Spotted-orchid [h]
			lf. spots v. small; fls. deep red/purple; lower lip [i]	*Dactylorhiza purpurella** Northern Marsh-orchid [i]
			fls. pink to white, spotted; lower lip [j]	*Dactylorhiza maculata* Heath Spotted-orchid [j]
		lvs. yellow-green, not spotted, erect	stem v. hollow; lf. tip hooded [f]	*Dactylorhiza incarnata* Early Marsh-orchid [f]
			corolla lip w. fine dots & lines [g]; mostly in S. & E.	*Dactylorhiza praetermissa* Southern Marsh-orchid [g]
			lip dark red/purple w. heavier markings [i]; N. & W.	*Dactylorhiza purpurella** Northern Marsh-orchid [i]

a   b   c   d   e   f   g   lf.   h   i   j

†Plants of the Genus *Dactylorhiza*, besides providing some of our commonest Orchids, also show considerable variation and hybridize freely.

Fls. pure white, or cream, erect; plant 15–60 cm high	bracts mostly longer than fls.; lvs. ovate-lanceolate		*Cephalanthera damasonium* White Helleborine
	bracts much shorter than fls.; lvs. lanceolate		*Cephalanthera longifolia* Narrow-leaved Helleborine [d]
Lower lip 3-lobed, yellow-green [d]	calcareous grass in S.		*Herminium monorchis** Musk Orchid [e]
Fls. purplish brown & white, veined	lip w. frilly sides [e]	marsh plant	*Epipactis palustris* Marsh Helleborine [a]
Stem downy near top	lvs. in 2 rows; fls. dark red [a]	on limestone	*Epipactis atrorubens** Dark-red Helleborine [b]
		lvs. lanceolate; plants often in clusters; fls. pale green/pink [b]	*Epipactis purpurata* Violet Helleborine [c]
	end of lip usu. turned under; stem often purplish	lvs. ovate; plants 1–3 together; fls. green to purple [c]	*Epipactis helleborine* Broad-leaved Helleborine [f]
	lvs. in 2 rows; lip greenish, spreading [f]	fls. yellow/green	*Epipactis leptochila** Narrow-lipped Helleborine [g]
Stem hairless near top	lvs. in 2 rows; fls. drooping [g]	fls. pale green/pink	*Epipactis phyllanthes** Green-flowered Helleborine

— bract

a   b   c   d   e   f   g

ARACEAE Arum Family

Plants with tiny greenish flowers arranged closely in a spike.
In Lords-and-ladies the flowers are hidden inside the sheath below the hooded club, but appear as red berries in the Autumn.

Waterside plant w. stout linear lvs.	fl. spike dense, oblique; sepals 6	lf. has wrinkled edge [a], sweet-scented if crushed	*Acorus calamus* * Sweet-flag
Fl. spike inside & protruding from a lfy. sheath	top of spike yellow; lvs. w. pale midrib	lvs. appear by Dec.; near S. coasts	*Arum italicum* * Italian Lords-and-ladies
	top of spike usu. purple [b]; lvs. w. dark midrib	lvs. appear in Spring	*Arum maculatum* Lords-and-ladies

LEMNACEAE Duckweeds

Small to minute floating aquatics often carpeting the surface, and usually with hanging roots. Flowers are hardly ever seen.
Stem and leaf and not differentiated so the whole plant (except the roots) is properly called a thallus.
Two other similar-sized plants may be associated with the Duckweeds: the two liverworts whose floating forms are illustrated here.

Riccia fluitans

Ricciocarpus natans

Minute egg-shaped rootless plant	thallus up to 1 mm across	in S.	*Wolffia arrhiza** Rootless Duckweed
Thalli stalked or branched, translucent		[a]	*Lemna trisulca* Ivy-leaved Duckweed
Only one rootlet to each thallus	thallus swollen below, 3–5 mm across	[c]	*Lemna gibba* Fat Duckweed
	thallus flat, up to 4 mm across	[d]	*Lemna minor* Common Duckweed
Several rootlets to each thallus	5–8 mm across	[b]	*Lemna polyrhiza* Greater Duckweed

SPARGANIACEAE Bur-reeds

Aquatic plants with flowers in green globular heads, spiky when in fruit.
The male heads are above the female.
The leaves are linear and may be upright or floating.

Lvs. 2–6 mm wide	usu. only one male head	lf. sheaths flat	*Sparganium minimum** Least Bur-reed
	usu. several male heads	lf. sheaths inflated [a]	*Sparganium angustifolium** Floating Bur-reed
		lvs. keeled near base	*Sparganium emersum* Unbranched Bur-reed
Lvs. wider, usu. erect, 3-sided	inflor. branched; fl. heads sessile		*Sparganium erectum* Branched Bur-reed
	inflor. not branched; some fl. hds. stalked		*Sparganium emersum* Unbranched Bur-reed

TYPHACEAE Bulrushes

Tall, erect waterside plants.
The inflorescence is a large dense spike, brown when ripe.
The male part is above the female.

| Lvs. 10–18 mm wide | male spike touches top of female spike | *Typha latifolia*
Bulrush (Reedmace) |
| Lvs. usu. under 5 mm wide | male spikes separated from female spike | *Typha angustifolia*
Lesser Bulrush |

CYPERACEAE Sedge Family

Sedges have grass-like leaves but solid stems (except *Cladium*) which are often 3-sided and do not have any cross-joints. There is a ligule where the leaf joins the stem.
The flowers are arranged in brown/green spikes which may be cylindrical, ovoid, or spherical, upright or drooping. A spike may consist of one or more spikelets.
Each spikelet contains several to many glumes (tiny bracts) behind which are the flowers.
The flowers may be male (with 2–3 stamens), female (with 2–3 stigmas) or both.
The measurements for female spikes are for ripe fruiting ones.
For identification fruiting plants are often more useful than flowering ones.
A larger number of uncommon species than usual has been omitted from the genus *Carex* (true sedges). The fifty species most likely to be found have been keyed out.

Types of inflorescence

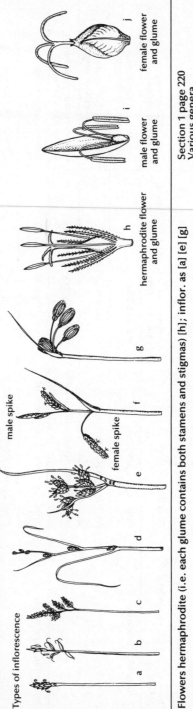

a   b   c   d   e   f   g   hermaphrodite flower and glume   h   male flower and glume   i   female flower and glume   j

male spike

female spike

Flowers hermaphrodite (i.e. each glume contains both stamens and stigmas) [h]; inflor. as [a] [e] [g]	Section 1 page 220 Various genera
Plant w. only one spike of flowers [b]	Section 2 page 223 *Carex*
Upper (male) spikes obvious different from lower (female) ones; [f]	Section 3 page 224 *Carex*
Plant w. several spikes, or spikelets, all similar in appearance; [c] & [d]	Section 4 page 228 *Carex*

CYPERACEAE Section 1

Each glume contains stamens & stigmas [a]

a

Stem hollow; lvs. saw-edged w. spine at tip	70–300 cm high	common in E. Anglia & W. Ireland, rare elsewhere	*Cladium mariscus* Great Fen-sedge
Fr. head like cotton wool	only 1 spike per stem	moors & bogs	*Eriophorum vaginatum* Hare's-tail Cottongrass
	lvs. flat; spike stalk rough	in fens, not acid bogs	*Eriophorum latifolium* Broad-leaved Cottongrass
	lvs. grooved; spike stalks smooth	in acid boggy places	*Eriophorum angustifolium* Common Cottongrass
One terminal spike only	spike consists of a fat cluster of spikelets	inflor. blackish [b]	*Schoenus nigricans* Black Bog-rush
	water plant w. branched lfy stem		*Eleogiton fluitans* Floating Club-rush
	stem w. 1–3 conspicuous inflated veined sheaths	spike in fr. like tuft of cotton wool	*Eriophorum vaginatum* Hare's-tail Cottongrass
	top sheath on stem w. a minute lf. or bract	a few minute bristles round seed; common; densely tufted [c]	*Trichophorum cespitosum* Deergrass
		delicate plant usu. under 15 cm [d]	*Isolepis cernua* Slender Club-rush

b

c

d

lowest glume at least ½ length of spike	stem 4-angled; spike 3–4 mm; poolsides etc.	*Eleocharis acicularis* Needle Spike-rush
	stem not angled; spike 5–7 mm; damp peaty places	*Eleocharis quinqueflora* Few-flowered Spike-rush
tufted plant; 3 stigmas per fl. [a]		[a] *Eleocharis multicaulis* Many-stalked Spike-rush
lowest glume almost encircles spike [b]	uncommon marsh plant; 2 stigmas per fl.	*Eleocharis uniglumis* Slender Spike-rush [b]
lowest glume about ½ encircles spike	common in pools; 2 stigmas per fl.	*Eleocharis palustris* Common Spike-rush
Inflor. w. several long flat or keeled lfy. bracts — spikes flattened	50–100 cm high	[c] *Cyperus longus** Galingale
	up to 20 cm high; on mud; in S. only	*Cyperus fuscus** Brown Galingale
lvs. w. rough edges	spikelets 10–20 mm long; stem sharply angled	[d, f] *Scirpus maritimus* Sea Club-rush
	spikelets about 4 mm, in a large, loose inflor.	[g] *Scirpus sylvaticus* Wood Club-rush
spike stalks smooth; lvs. grooved	in acid boggy places	[e] *Eriophorum angustifolium* Common Cottongrass
spike stalks rough; lvs. flat	in fens, not acid bogs	*Eriophorum latifolium* Broad-leaved Cottongrass

glumes

in fr.

221

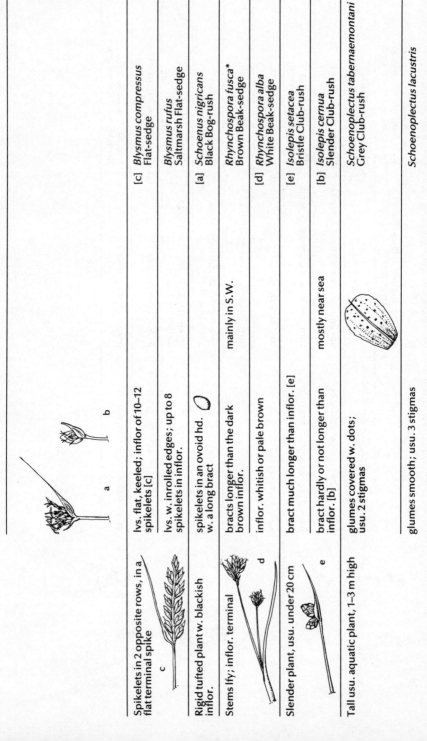

Spikelets in 2 opposite rows, in a flat terminal spike	lvs. flat, keeled; inflor of 10–12 spikelets [c]	[c] *Blysmus compressus* Flat-sedge
	lvs. w. inrolled edges; up to 8 spikelets in inflor.	*Blysmus rufus* Saltmarsh Flat-sedge
Rigid tufted plant w. blackish inflor.	spikelets in an ovoid hd. w. a long bract	[a] *Schoenus nigricans* Black Bog-rush
Stems lfy; inflor. terminal	bracts longer than the dark brown inflor. — mainly in S.W.	*Rhynchospora fusca** Brown Beak-sedge
	inflor. whitish or pale brown	[d] *Rhynchospora alba* White Beak-sedge
Slender plant, usu. under 20 cm	bract much longer than inflor. [e]	[e] *Isolepis setacea* Bristle Club-rush
	bract hardly or not longer than inflor. [b] — mostly near sea	[b] *Isolepis cernua* Slender Club-rush
Tall usu. aquatic plant, 1–3 m high	glumes covered w. dots; usu. 2 stigmas	*Schoenoplectus tabernaemontani* Grey Club-rush
	glumes smooth; usu. 3 stigmas	*Schoenoplectus lacustris*

CYPERACEAE Section 2 Stems with only one flower spike.
Each glume with either stamens or stigmas, but not both.

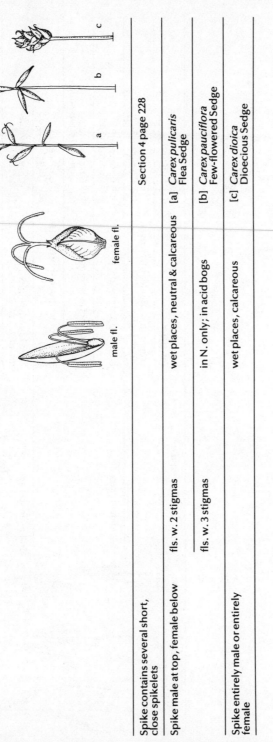

male fl.   female fl.

Spike contains several short, close spikelets		Section 4 page 228
Spike male at top, female below	fls. w. 2 stigmas	wet places, neutral & calcareous	[a] *Carex pulicaris* Flea Sedge
	fls. w. 3 stigmas	in N. only; in acid bogs	[b] *Carex pauciflora* Few-flowered Sedge
Spike entirely male or entirely female		wet places, calcareous	[c] *Carex dioica* Dioecious Sedge

223

CYPERACEAE Section 3 Male and female spikes obviously different

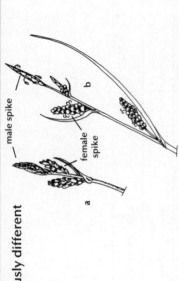

male spike

female spike

a

b

Female fls. w. 2 stigmas; glumes usually blackish

usu. only 1 male spike

lowest bract shorter than inflor.; mt. plant	*Carex bigelowii* Stiff Sedge	[a]
common, variable; often over 30 cm high	*Carex nigra* Common Sedge	[b]
1–2 male spikes		
lvs. up to 3 mm wide	*Carex nigra* Common Sedge	[c]
lowest bract as long as inflor.	*Carex acuta* Slender Tufted-sedge	[d]
lowest bract much shorter than inflor.	*Carex elata* Tufted-sedge	[e]
2–4 male spikes		
stem bluntly angled; by water in mt. districts	*Carex aquatilis* Water Sedge	[f]
stem sharply angled	*Carex acuta* Slender Tufted-sedge	[d & g]

f

g

male

female

male

female

female

male

female

c

d

e

Fr. hairy or downy

			Species
Lf. sheaths w. long hairs			*Carex hirta* Hairy Sedge
usu. 2–3 male spikes; female ones well-spaced	50–120 cm high; fr. densely hairy		[a] *Carex lasiocarpa* Slender Sage
	10–40 cm high; glaucus; fr. minutely downy		[b] *Carex flacca* Glaucus Sedge
male spike v. thin; female ones almost round	10–30 cm high heathlands etc.		[c] *Carex pilulifera* Pill Sedge
	usu. under 15 cm; in dry grassland [d]		[d] *Carex caryophyllea* Spring-sedge

Usu. 2 or more male spikes

			Species
4–6 male spikes	by water edge		*Carex riparia* Greater Pond-sedge
lvs. bluish, 2–4 mm wide; fr. blunt [e]	10–40 cm high		[e] *Carex flacca* Glaucus Sedge
stem sharply angled even near base; ligule acute	lvs. 4–5 mm wide; fr. clearly longer than glume		[f] *Carex vesicaria* Bladder-sedge
	lvs. 7–10 mm wide; fr. hardly longer than glume		[g] *Carex acutiformis* Lesser Pond-sedge
male glumes pale brown; fr. acutely pointed [h]	ligule rounded		[h] *Carex rostrata* Bottle Sedge

Female spikes 5–16 cm long, drooping	tall plant of shady places, 60–150 cm high		*Carex pendula* Pendulous Sedge
Female spikes usu. drooping, 2–5 cm long	stem w. v. sharp, rough angles fr. [a]	by water	[a] *Carex pseudocyperus* Cyperus Sedge
	female spikes loose, 3–4 mm wide fr. [b]	woodlands	[b] *Carex sylvatica* Wood-sedge
	lvs. & female spikes 5–10 mm wide fr. [c]	marshy places	[c] *Carex laevigata* Smooth-stalked Sedge
Most female spikes clustered near top of stem	stem sharply angled, rough; lvs. sl. hairy below	male spike almost hidden by female spikes	[d] *Carex pallescens* Pale Sedge
	male spike stalked	20–50 cm high; most ripe fr. bent down; calcareous soils	[h] *Carex lepidocarpa* Long-stalked Yellow-sedge
		10–20 cm high; on acid soils	[e] *Carex demissa* Common Yellow-sedge
		20–40 cm high; bracts v. long, narrow	[f] *Carex extensa* Long-bracted Sedge
		saltmarshes	
		usu. 5–15 cm high	[g] *Carex serotina* Small-fruited Yellow-sedge

a   b   c

male
female
h

male
female
d

e

male
female
f

g

Fr. rounded (not beaked) at top

stem sharply angled, rough; fr. ribbed [a]	usu. in pools	[a] *Carex limosa* Bog-sedge
stem smooth; lvs. bluish fr. [b]	damp grassy places	[b] *Carex panicea* Carnation Sedge

Most female spikes over twice as long as wide

female glumes when ripe usu. dark brown	female glumes blunt w. tiny point	[c] *Carex binervis* Green-ribbed Sedge
	female glumes acute	[d] *Carex hostiana* Tawny Sedge
female glumes green/brown	often coastal	*Carex distans* Distant Sedge

Lowest bract reaches at least up to male spike

some fr. bent down [e]	calcareous soils; 20–50 cm high	[e] *Carex lepidocarpa* Long-stalked Yellow-sedge
fr. w. straight beak [f]	acid soils; 10–20 cm high	[f] *Carex demissa* Common Yellow-sedge

male spike

female spikes

glumes

a b c d e f

Female spikes shortly oblong

female glumes when ripe usu. dark brown [d]		[d] *Carex hostiana* Tawny Sedge

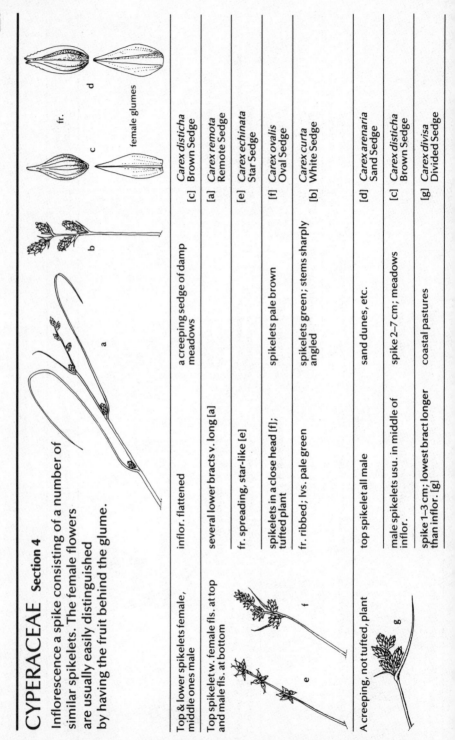

CYPERACEAE Section 4

Inflorescence a spike consisting of a number of
similar spikelets. The female flowers
are usually easily distinguished
by having the fruit behind the glume.

fr.

female glumes

Top & lower spikelets female, middle ones male	inflor. flattened	a creeping sedge of damp meadows	[c] *Carex disticha* Brown Sedge
Top spikelet w. female fls. at top and male fls. at bottom	several lower bracts v. long [a]		[a] *Carex remota* Remote Sedge
	fr. spreading, star-like [e]		[e] *Carex echinata* Star Sedge
	spikelets in a close head [f]; tufted plant	spikelets pale brown	[f] *Carex ovalis* Oval Sedge
	fr. ribbed; lvs. pale green	spikelets green; stems sharply angled	[b] *Carex curta* White Sedge
A creeping, not tufted, plant	top spikelet all male	sand dunes, etc.	[d] *Carex arenaria* Sand Sedge
	male spikelets usu. in middle of inflor.	spike 2–7 cm; meadows	[c] *Carex disticha* Brown Sedge
	spike 1–3 cm; lowest bract longer than inflor. [g]	coastal pastures	[g] *Carex divisa* Divided Sedge

Inflorescences and female glumes

Stout tufted plant, 25–150 cm high	stem slender; lvs. up to 3 mm wide		[a] *Carex divulsa* Grey Sedge
	lvs. dark green; inflor. w. branches	in large tussocks up to 1.5 m high	[b] *Carex paniculata* Tussock Sedge
	lvs. bright green; spikelets stalkless	up to 100 cm high; stems sharply angled	[c] *Carex otrubae* False Fox-sedge
Inflor. 5–15 cm; lower spikelets well-spaced [a]			[a] *Carex divulsa* Grey Sedge
Lvs. 4–10 mm wide	stems stout, sharply angled		[c] *Carex otrubae* False Fox-sedge
Lowest bract at least as long as inflor. [d]	stems slender	usu. in coastal pastures	[d] *Carex divisa* Divided Sedge
Bracts & glumes tinged purple/red	stems rough; ripe fr. greenish	usu. in damp, grassy places	[e] *Carex spicata* Spiked Sedge
Spike mid to dark brown	lowest bract v. short [f]	in damp places	[f] *Carex diandra* Lesser Tussock-sedge
Bracts green/brown		in dry grassland	[e] *Carex muricata* Prickly Sedge

GRAMINEAE Grasses

Grass flowers are arranged in spikelets, each spikelet consisting of two small bracts (only 1 in *Lolium*) (glumes) at the base, and from one to about a dozen florets above these. [See diag. i.]

Each floret consists of two more bracts (a lemma and a palea) which enclose the 3 stamens, 2 stigmas and ovary [see diag j].

The stems are usually hollow, with swollen joints (nodes).

The leaves form a sheath at their base and bear a ligule [diag. m] at their point of divergence from the stem.

Occasionally the ligule is only a ring of hairs or may be missing entirely.

NOTE: If the plant being examined has solid or 3-sided stems, or 3 stigmas (or none) to each flower or drooping catkin-like spikes, then it is likely to be a Sedge (p. 219). If each flower has 6 sepals then it is probably a Rush (p. 206).

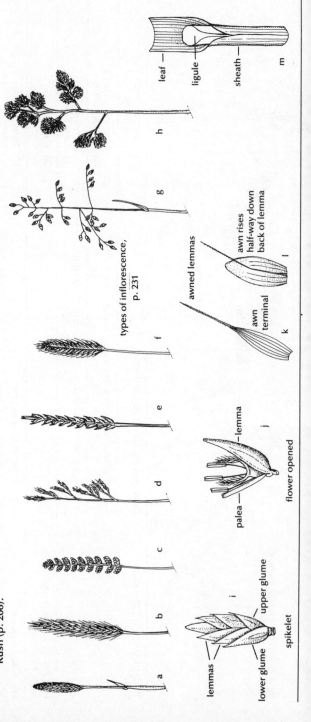

a

b

c

d

e

f

types of inflorescence, p. 231

g

h

spikelet

lemmas

lower glume

upper glume

i

palea

lemma

flower opened

j

awn terminal

k

awned lemmas

awn rises half-way down back of lemma

l

leaf

ligule

sheath

m

GRAMINEAE Key to Sections

NOTE: The diagrams referred to are all on the previous page.

Ligule consists only of a ring of hairs			Section 1 page 232
Inflor. a single spike which may be cylindrical or flat or 1-sided; spikelets almost stalkless, on main stem or on tightly packed branchlets [a–f]	spike approximately cylindrical; spikelets set all round stem [a, b]	spikelets long-awned or spike whiskery [b]	Section 2 page 233
		spike not whiskered [a]	Section 3 page 235
Beware of an inflorescence which is closed up and looks like a spike	spike 1-sided; stem visible at back [c–d]		Section 4 page 236
	spikelets alternate on opposite sides of stem [e–f]		Section 5 page 237
Inflor. open or loose or spreading or drooping [g–h]	florets w. an obvious awn usu. 3 mm or more long [k–l]	awn rises from or near tip of lemma [k]	Section 6 page 240
		awn rises ½-way or more down back of lemma [l]	Section 7 page 242
	spikelets w. only 1 or 2 florets inside the glumes		Section 8 page 244
	spikelets usu w. 3–10 florets [i]		Section 9 page 247

231

GRAMINEAE Section 1 Ligule consists merely of a ring of hairs

2–3 m high; lvs. 10–20 mm wide		in wet places or shallow water	*Phragmites australis* Common Reed
Inflor. of 3–5 spikes in a terminal umbel [b]		sandy places, usu. near the sea	*Cynodon dactylon** Bermuda-grass
Inflor. a tight spike w. many long bristles [c]			*Setaria viridis** Green Bristle-grass
In saltmarshes; spikes erect [d]; spikelets w. 1 floret	inflor. 12–40 cm long of 3–6 spikes	anthers 8–13 mm [a + d]	*Spartina anglica* Common Cord-grass
		anthers 5–8 mm	*Spartina × townsendii* Townsend's Cord-grass
	inflor. 4–10 cm long of 2–3 spikes		*Spartina maritima* Small Cord-grass
Inflor. w. under 10 spikelets	10–40 cm high ligule [e]	heathlands	*Danthonia decumbens* Heath-grass
Spikelets often purplish; stem swollen in Autumn	often forms tussocks 30–120 cm high; ligule [f]	damp heaths etc.	*Molinia caerulea* Purple Moor-grass

GRAMINEAE Section 2 Inflorescence a single more-or-less cylindrical spike. Glumes and/or lemmas awned.

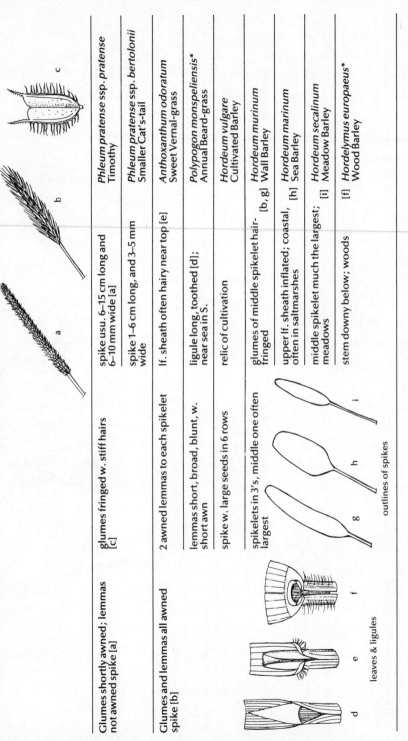

Glumes shortly awned; lemmas not awned spike [a]	glumes fringed w. stiff hairs [c]		spike usu. 6–15 cm long and 6–10 mm wide [a]	*Phleum pratense* ssp. *pratense* Timothy
			spike 1–6 cm long, and 3–5 mm wide	*Phleum pratense* ssp. *bertolonii* Smaller Cat's-tail
Glumes and lemmas all awned spike [b]	2 awned lemmas to each spikelet		lf. sheath often hairy near top [e]	*Anthoxanthum odoratum* Sweet Vernal-grass
	lemmas short, broad, blunt, w. short awn		ligule long, toothed [d]; near sea in S.	*Polypogon monspeliensis** Annual Beard-grass
	spike w. large seeds in 6 rows		relic of cultivation	*Hordeum vulgare* Cultivated Barley
	spikelets in 3's, middle one often largest		glumes of middle spikelet hair-fringed [b, g]	*Hordeum murinum* Wall Barley
			upper lf. sheath inflated; coastal, often in saltmarshes [h]	*Hordeum marinum* Sea Barley
			middle spikelet much the largest; meadows [i]	*Hordeum secalinum* Meadow Barley
			stem downy below; woods [f]	*Hordelymus europaeus** Wood Barley

outlines of spikes

leaves & ligules

233

Lf. sheath w. ring of hairs round top [a]	spike loose; 2 awns to each spikelet		Anthoxanthum odoratum Sweet Vernal-grass
Stems w. bulbs at base		saltmarshes	Alopecurus bulbosus* Bulbous Foxtail
Stems sharply bent at lower nodes [b]	glumes clearly joined to each other [d]	in arable land	Alopecurus myosuroides Black-grass
	glumes almost free from each other	in wet places	Alopecurus geniculatus Marsh Foxtail
Under 15 cm high	spike loose, up to 5 cm long	sandy places	Aira praecox Early Hair-grass
Spike 5–10 mm wide; ligule v. short [c]		meadows, etc.; v. common	Alopecurus pratensis Meadow Foxtail
Spike 3–6 mm wide	glumes clearly joined to each other [d]	in arable land	Alopecurus myosuroides Black-grass

a

b

c

d

GRAMINAE Section 3 Inflorescence a single more or less cylindrical spike, not whiskery.

Spikelets not or hardly awned, set all round the stem.

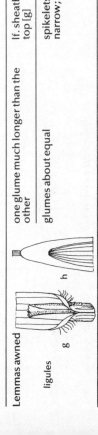

a

Tall grass w. spikes mostly over 8 cm long	stout plant of sand dunes	ligules long, narrow [c]	*Ammophila arenaria* Marram
		ligule v. short [d]	*Leymus arenarius* Lyme-grass
	glumes fringed w. stiff hairs	cultivated and naturalized glumes [f]	*Phleum pratense* ssp. *pratense* Timothy
Glumes fringed w. stiff spreading hairs	on sand dunes; usu. under 16 cm high [a]	glumes tapered to a point [e]	*Phleum arenarium* Sand Cat's-tail
	spike 6–15 cm long and 6–10 mm wide	glumes w. v. short awns; cultivated and naturalized [f]	*Phleum pratense* ssp. *pratense* Timothy
	spike 1–6 cm long and 3–5 mm wide	glumes w. v. short awns; native; grassland	*Phleum pratense* ssp. *bertolonii* Smaller Cat's-tail
Lemmas awned	one glume much longer than the other	lf. sheath w. ring of hairs round top [g]	*Anthoxanthum odoratum* Sweet Vernal-grass
ligules	glumes about equal	spikelets often silvery; lvs. v. narrow; ligule [h]	*Aira praecox* Early Hair-grass

c

d

e

f

g

h

▶

235

Character	Habitat / notes	Species
Lvs. fine, usu. hairy	usu. in dry grassland	*Koeleria macrantha* Crested Hair-grass
Ligule extremely short	in N. only	*Sesleria albicans* Blue Moor-grass
Spike often conical [a]	a casual	*Phalaris canariensis* Canary-grass
Spike narrowly cylindrical	in wet places; ripe anthers orange	*Alopecurus aequalis** Orange Foxtail

GRAMINEAE Section 4 Inflorescence a one-sided spike

Lemmas long-awned, giving spike a whiskery appearance — ligule 3–10 mm long, triangular	inflor. short, broad [b]	*Cynosurus echinatus** Rough Dog's-tail
glumes: smaller glumes at least ½ as long as other [c]	common	*Vulpia bromoides* Squirreltail Fescue
one glume awned, 3 mm or more, other almost nil [d]	sand dunes	*Vulpia fasciculata** Dune Fescue
smaller glume at least ¼ as long as other [e]		*Vulpia myuros* Rat's-tail Fescue
smaller glume minute [f]; plant often purplish	sand dunes	*Vulpia ciliata* ssp. *ambigua** Bearded Fescue
Spikelets fat, often drooping; 2–3 fl'd	shady places, usu. on limestone	*Melica nutans** Mountain Melick
Spike dense, almost cylindrical		*Cynosurus cristatus* Crested Dog's-tail

Spikelets w. 3 or more florets		
spikelets in 2 rows, mostly stalkless; spike [a]	coastal	*Desmazeria marina* Sea Fern-grass
glumes blunt, containing 3–5 florets	inflor. usu. spreading; saltmarshes, etc.	*Puccinellia rupestris** Stiff Saltmarsh-grass
glumes acute, containing 3–10 florets inflor [b]	dry places	*Desmazeria rigida* Fern-grass
Spikelets w. one floret		
lvs. v. fine; in dense tufts	moorlands, etc.	*Nardus stricta* Mat-grass

GRAMINEAE Section 5 Inflorescence a single spike. Spikelets arranged alternately or in alternate clusters on opposite sides of the stem.

Inflor. dense w. long awns; spikelets in 3's, middle one often fatter than the others		
glumes clearly arranged in 6 rows; relic of cultivation	2 rows of glumes fat, 4 rows thin	Hordeum distichon Two-rowed Barley
	all 6 rows fat w. seeds	*Hordeum vulgare* Six-rowed Barley
glumes of middle spikelet hair-fringed	common; spike shape [c]	*Hordeum murinum* Wall-Barley
upper lf. sheath inflated	saltmarshes; spike shape [d]	*Hordeum marinum* Sea Barley
middle spikelet much the largest	meadows; [d] spike shape [e]	*Hordeum secalinum* Meadow Barley
stem downy below [f]	woods	*Hordelymus europaeus** Wood Barley

leaf & ligule

outlines of spikes

c d e f

▶

Lemmas and/or glumes awned

spikelets w. narrow edge to stem [a]; only 1 glume			*Lolium perenne* ssp. *multiflorum* Italian Rye-grass
usu. 8 or more florets to each spikelet [a]	awn at least as long as lemma; plant downy [b]		*Brachypodium sylvaticum* False Brome
	awn shorter than lemma; plant mostly hairless [d]		*Brachypodium pinnatum* Tor-grass
lvs. w. inrolled edges	maritime plant, often bluish grey		*Elymus pycnanthus* Sea Couch
awn often longer than lemma	glumes narrow, tapered; nodes may be finely hairy [c]		*Elymus caninus* Bearded Couch
	glumes & lemmas broad; relic of cultivation		*Triticum aestivum* Wheat
awn usu. shorter than lemma [e]	common weed; stem and nodes hairless		*Elymus repens* Common Couch

Spike v. narrow; spikelet w. 1 floret

stems usu. curved and under 10 cm high	S. & E. coasts	*Parapholis incurva* Curved Hard-grass
stems usu. straight and over 15 cm high [f]	coastal	*Parapholis strigosa* Hard-grass

spikelet

Usu. under 20 cm high		coastal; ligule [e]	*Desmazeria marina* Sea Fern-grass
Spikelets broadside to the stem	spikelets usu. in pairs; spike usu. over 15 cm	robust bluish sand dune plant	*Leymus arenarius* Lyme-grass
	creeping, often prostrate; in sand dunes	lf. ribs minutely downy spikelet [a]; ligule [b]	*Elymus farctus* Sand Couch
	lvs. w. inrolled edges & prominent ribs	saltmarshes & sand dunes ligule & auricles [c]	*Elymus pycnanthus* Sea Couch
	glumes lanceolate	common weed [f] ligule & auricles [d]	*Elymus repens* Common Couch
	glumes & lemmas broad, oblong	relic of cultivation	*Tricticum aestivum* Wheat
Mostly spikelets w. 1 glume		ligule [f]	*Lolium perenne* Perennial Rye-grass

auricles at lf. base

ligules

GRAMINEAE Section 6

Inflorescence loose or loosely clustered or spreading, lemmas with a terminal awn usually 3 mm or more long.

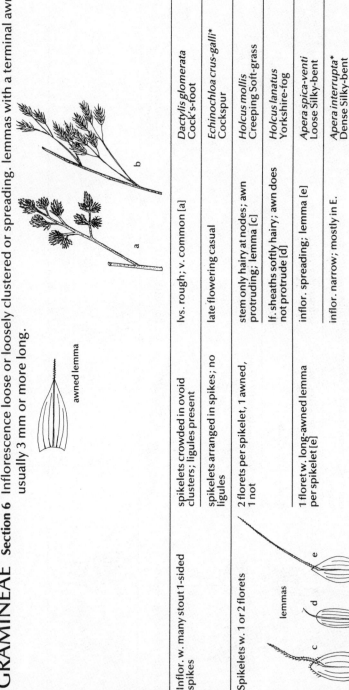

awned lemma

lemmas

a

b

c d e

Inflor. w. many stout 1-sided spikes	spikelets crowded in ovoid clusters; ligules present	lvs. rough; v. common [a]	*Dactylis glomerata* Cock's-foot
	spikelets arranged in spikes; no ligules	late flowering casual	*Echinochloa crus-galli** Cockspur
Spikelets w. 1 or 2 florets	2 florets per spikelet, 1 awned, 1 not	stem only hairy at nodes; awn protruding; lemma [c]	*Holcus mollis* Creeping Soft-grass
		lf. sheaths softly hairy; awn does not protrude [d]	*Holcus lanatus* Yorkshire-fog
	1 floret w. long-awned lemma per spikelet [e]	inflor. spreading; lemma [e]	*Apera spica-venti* Loose Silky-bent
		inflor. narrow; mostly in E.	*Apera interrupta** Dense Silky-bent
Basal lvs. either tightly rolled or v. fine	stem lvs. flat, 2–4 mm wide	woods	*Festuca heterophylla** Various-leaved Fescue

If. sheath split to its base	ligule w. small auricle [a]	*Festuca longifolia* Hard Fescue
If. sheath closed like a tube when young	v. common; ligule [b]	*Festuca rubra* Red Fescue
Nodes purple; plant hairless		
awn longer than lemma	shady places inflor [b] on p. 240	*Festuca gigantea* Giant Fescue
awn shorter than lemma	branches long, drooping; sheaths long haired [f]	*Bromuc ramosus* Hairy-brome
Lower glume w. 1 nerve, upper w. 3 nerves; both narrowly acute	spikelets wider at top [e]; awn nearly as long as lemma	*Bromus madritensis** Compact Brome
	awn much shorter than lemma inflor. [d] sheath & ligule [g]	*Bromus erectus* Upright Brome
inflor. branches long, drooping	spikelets 4–6 cm, including 15–30 mm awns [c]; common	*Bromus sterilis* Barren Brome
	spikelets 7–9 cm, including 3–6 cm awns; rare	*Bromus diandrus** Great Brome
inflor. erect [e]		*Bromus madritensis** Compact Brome

ligules

f   g   ligules

a   b   ligules

c   d   e

▶

Glumes & lemmas narrowly lanceolate	spikelets drooping, 7–9 cm, including awns	awns 3–6 cm long	*Bromus diandrus** Great Brome
Lower lf. sheaths hardly hairy	inflor. often drooping to one side		*Bromus secalinus** Rye Brome
Spikelets downy; ligule hairy	some spikelets longer than their stalks	inflor. usu. erect [a]	*Bromus hordeaceus* agg. Soft-brome
Spikelets almost hairless	spikelets mostly longer than their stalks	tip of grain visible inside lemma inflor. [c]	*Bromus lepidus* Slender Soft-brome
		grain not visible; inflor. [a]	*Bromus hordeaceus* agg. Soft-brome
	spikelets 18–28 mm, on long stalks	inflor. drooping when ripe [b]	*Bromus commutatus* Meadow Brome
	spikelets 12–16 mm	inflor. erect	*Bromus racemosus* Smooth Brome

a   b   c

GRAMINEAE Section 7

Inflorescence loose or loosely clustered or spreading. Lemmas with an awn which arises from the back of the lemma (often about half-way down) not from the tip.

d

Each spikelet w. only 1 awn	spikelet contains 2 lemmas, one awnless	spikelets 17–20 mm long; relic of cultivation	*Avena sativa* Oat
		spikelet 7–10 mm long; v. common; lemma & awn [d]	*Arrhenatherum elatius* False Oat-grass

		Species
lvs. v. fine, bristle-like	in dense tufts; on heaths in S. & W. only	*Agrostis curtisii* Bristle Bent
lvs. fine but flat	widespread	*Agrostis canina* Brown Bent
Spikelet w. 1 long & 1 short awn	longer awned lemma [a]	*Arrhenatherum elatius* False Oat-grass
Lvs. edged w. tiny sharp teeth	often in large tussocks, 30–200 cm high	damp places	*Deschampsia cespitosa* Tufted Hair-grass
Under 15 cm high	inflor. compact, almost spike-like [b]	sandy places; in fl. April–May	*Aira praecox* Early Hair-grass
	inflor. loose, silvery [c]	sandy places	*Aira caryophyllea* Silver Hair-grass
Spikelets w. awns up to 7 mm long	lvs. bristle-fine; spikelets 2-awned	inflor. branches often wavy; 25–100 cm high	*Deschampsia flexuosa* Wavy Hair-grass
		10–30 cm high; inflor. silvery [c]	*Aira caryophyllea* Silver Hair-grass
	lvs. flat; spikelets w. 2–4 awns; ligule up to 2 mm [d]	inflor. yellowish green	*Trisetum flavescens* Yellow Oat-grass

▶

Awn 25 mm or more long	lemmas w. tuft of hairs at the base [a]	all lemmas w. an oval scar at base inside [a]	*Avena fatua* Common Wild-oat
		only lowest lemma has a basal scar	*Avena sterilis* Winter Wild-oat
	lemmas w'out basal tuft of hairs	lemma tipped w. bristles 3 mm or more long [b]	*Avena strigosa* Bristle Oat
		lemma only toothed at tip; cultivated	*Avena sativa* Oat
lemmas tipped w.	spikelets over 15 mm long	lemma & awn [b] ligule [c]	*Avena strigosa* Bristle Oat
	spikelets 5–7 mm long, often yellowish	ligule v. short [d]	*Trisetum flavescens* Yellow Oat-grass
Lowest lf. sheath hairy; spikelets w. 2–3 awns			*Avenula pubescens* Downy Oat-grass
l.f. sheaths almost hairless; spikelets w. 3–6 awns			*Avenula pratense* Meadow Oat-grass

lemmas

a   b

c   d

GRAMINEAE   Section 8

Inflorescence loose or loosely clustered or spreading. Spikelets containing 1 or 2 lemmas not obviously awned, but a close look (use a lens) may reveal a fine or a short awn.

Spikelets arranged in spikes; no ligules	a late flowering casual	*Echinochloa crus-galli** Cockspur

Spikelets w. 2 florets			
each lemma w. v. fine awn from near base [b]	large tufted plant, 50–200 cm of damp places; lvs. v. rough		*Deschampsia cespitosa* Tufted Hair-grass
1 lemma w. tiny often curved awn at tip [b]	v. common		*Holcus lanatus* Yorkshire-fog
inflor. branches in clusters of 3 or more	lemmas v. blunt, much longer than glumes [c, f]		*Catabrosa aquatica* Whorl-grass
	ligule shorter than its diameter [d]; sheaths smooth; spikelet [g]		*Poa nemoralis* Wood Meadow-grass
	ligule long [e]; lf. sheaths rough		*Poa trivialis* Rough Meadow-grass
branches often in 2's [h]; spikelets few, broad			*Melica uniflora* Wood Melick
Spikelets w. many silky hairs surrounding the lemmas			
lvs. downy on upper side; ligule 2–5 mm long [i]	spikelet hairs a little longer than lemma		*Calamagrostis canescens* Purple Small-reed
lvs. hairless above; ligule 4–12 mm long [j]	spikelet hairs about twice as long as lemma [k]		*Calamagrostis epigejos* Wood Small-reed
Spikelets in dense heavy clusters [l]	60–120 cm high; lvs. mostly 8–14 mm wide	wet places	*Phalaris arundinacea* Reed Canary-grass
Lemmas much longer than the broad glumes [m]		wet places	*Catabrosa aquatica* Whorl-grass

lemmas a b c; ligules d e; h; i j ligules; k; l; m

Glumes w. 3 or 5 nerves	inflor. branches in clusters of 3 or more	ligule longer than its width; lvs. 5–10 mm wide inflor. [a]
		ligule v. short; lvs. 2–3 mm wide inflor.
	branches often in 2's [b]	spikelets few, broad, brownish
Ligule longer than its width [c, d]	inflor closes up after flowering	spreads by horizontal leafy stems above ground; ligule [c]
	inflor. remains open	spreads by underground stems ligule [d]
Ligule v. short [e]	inflor. often brownish to purplish	

Milium effusum Wood Millet

Poa nemoralis Wood Meadow-grass

Melica uniflora Wood Melick

Agrostis stolonifera Creeping Bent

Agrostis gigantea Black Bent

Agrostis capillaris Common Bent

ligules

GRAMINEAE Section 9 Inflorescence loose or loosely clustered, or spreading. Spikelets with 3 or more florets. Lemmas not awned.

a

b

Only 3 (or even 2) florets per spikelet	spikelets packed in dense ovoid clusters [c]	lvs. v. rough	*Dactylis glomerata* Cock's-foot
	glumes & lemmas blunt; in wet places		*Catabrosa aquatica* Whorl-grass
	tufts of tiny lvs. grow from spikes [d]	mts.	*Festuca vivipara* Viviparous Fescue
	ligule v. short	inflor. delicate, often nodding; shady places [a]; spikelet [a]	*Poa nemoralis* Wood Meadow-grass
		inflor. usu. erect or spreading [e]	*Poa pratensis* Smooth Meadow-grass
	ligule long e.g. [f]	inflor. branches 1–2 together, 5–30 cm high	*Poa annua* Annual Meadow-grass
		lf. sheaths slightly rough; common ligule [f]	*Poa trivialis* Rough Meadow-grass
		lf. sheaths smooth; 30–150 cm high; wet places	*Poa palustris** Swamp Meadow-grass
Spikelets about as wide as long, hanging down in 1's [b]	glumes & lemmas v. blunt		*Briza media* Quaking-grass

lemma

glume

c

d

e

f

▶

Spikelets packed in dense ovoid clusters; [c] on previous page	ligule long; lemmas bristle-topped	stout, rough, tufted	*Dactylis glomerata* Cock's-foot
Inflor. stiff, often 1-sided; branches	glumes blunt w. 3–5 lemmas [c]	saltmarshes, etc.	*Puccinellia rupestris** Stiff Saltmarsh-grass
	glumes acute w. up to 10 lemmas [b]	up to 20 cm high, in dry places [a]	*Desmazeria rigida* Fern-grass
Basal lvs. v. fine, almost bristle-like	Ligule extremely short; lemma acute or w. fine awn	tufts of tiny lvs. grow from spikelets [d]; mts.	*Festuca vivipara* Viviparous Fescue
		lemma acute but not awned [h]	*Festuca tenuifolia* Fine-leaved Sheep's Fescue
		young lf. sheaths closed almost to top [i]	*Festuca rubra* Red Fescue
		all lf. sheaths split almost to base [j]; spikelet [e]	*Festuca ovina* Sheep's-fescue
	ligule acute [g]; stems w. blub-like base	coastal sands in S. & E.	*Poa bulbosa** Bulbous Meadow-grass
	ligule blunt; spikelet [f]	dry grassland	*Poa angustifolia* Narrow-leaved Meadow-grass

Most spikelets 12 mm or more long	ligule v. short; auricles at lf. base [a]; inflor. branches in 2's	shorter branches w. only 1 or 2 spikelets	*Festuca pratensis* Meadow Fescue
		shorter branches w. 3 or more spikelets	*Festuca arundinacea* Tall Fescue
	tip of lemma clearly 3-toothed [c]; spikelet [b]	10–45 cm high	*Glyceria declinata* Small Sweet-grass
	lf. sheath minutely rough	lower branches of inflor. 2–5 together; ligule [e]	*Glyceria plicata* Plicate Sweet-grass
		inflor. branches 1–3 together	*Glyceria × pedicellata* Hybrid Sweet-grass
	lf. sheath smooth	inflor. branches in 1's or 2's; ligule [d]	*Glyceria fluitans* Floating Sweet-grass
		lower branches in 2's or 3's	*Glyceria × pedicellata* Hybrid Sweet-grass
Lemmas convex but not keeled on the back	saltmarsh plants; lemmas obtuse	lemma midrib reaches tip [f]; branches 2–4 together	*Puccinellia fasciculata** Borrer's Saltmarsh-grass
		inflor. branches up to 3 together lemma [g]	*Puccinellia maritima* Common Saltmarsh-grass
		inflor. branches bent down after flowering; lemma [h]	*Puccinellia distans* Reflexed Saltmarsh-grass
	under 30 cm high; v. common weed	ligule [i]	*Poa annua* Annual Meadow-grass
	lemma acute w. 3 or 5 nerves	lemmas 5-nerved [m]; lf. base w. auricles [j]	*Festuca pratensis* Meadow Fescue
		lemmas 3-nerved [n]; ligule [k]; in shady places	*Festuca altissima** Wood Fescue
	lemma obtuse; ligule w. a central point [l]; lemma [o]	60–200 cm high in or by freshwater	*Glyceria maxima* Reed Sweet-grass

249

▶

Ligule much shorter than its diameter, lemmas keeled	top lf. usu. longer than its sheath	*Poa nemoralis* Wood Meadow-grass
	inflor. delicate, often nodding; shady places	
	top lf. usu. shorter than its sheath; ligule [a]	*Poa pratensis* Smooth Meadow-grass
	inflor. usu. erect or spreading	

a

Lemmas keeled	stems much flattened; usu. 20–40 cm high	*Poa compressa* Flattened Meadow-grass
	dry places;	
	stem cross-section	
	spikelets usu. bear tufts of tiny lvs. [b]	*Poa alpina** Alpine Meadow-grass
	mts.	
	lf. sheaths slightly rough; 20–60 cm high	*Poa trivialis* Rough Meadow-grass
	inflor. branches in clusters of 3–7 ligule [d]	
	5–30 cm high	*Poa annua* Annual Meadow-grass
	inflor. branches 1 or 2 together [c]; ligule [e]	
	30–150 cm high	*Poa palustris** Swamp Meadow-grass
	wet places	

stem cross-section

b

c

d

e

Simplified keys

The following simplified keys are provided to the common plants of large groups which are often difficult to identify.

18 common Ferns   page 252
18 common white Umbellifers (like Cow Parsley)   page 255
25 common Labiates (Deadnettle type of flower)   page 257
15 common yellow Composites (Dandelion-like flowers)   page 260
5 common Docks   page 262

Ferns

Parsleys

Deadnettles

Dandelions

Docks

Simplified key to 18 common ferns
The full key with further drawings is on page 60.

SPECIAL TERMS USED FOR FERNS

Frond a whole fern leaf (and all its leaflets)
Pinna a leaflet springing from the main stem
Pinnule a subdivision, or lobe, of a pinna
Sorus a patch of spore cases, usually on the back of the frond
Indusium the cover over a sorus, best seen when young, as it may shrivel and fall off.

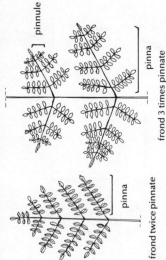

pinnule

pinna

pinna

frond 3 times pinnate

pinna

frond twice pinnate

c

Bracken

b

a

a pinnule with sori

indusium   indusium

attached at its centre   attached at its edge

d

e

Frond simple [a]	*Phyllitis scolopendrium*	Hart's-tongue
Frond once pinnate or deeply lobed	fronds covered underneath w. brown scales [d] — *Ceterach officinarum*	Rustyback
	sori distinct, more or less circular [e] — *Polypodium vulgare*	Polypody

(illustration)	Description	Name	Common name
a, b, c	frond lobes entire [a]	*Blechnum spicant*	Hard Fern
	stem black [b]	*Asplenium trichomanes*	Common Spleenwort
	stem brownish; fronds tough [c]	*Asplenium marinum*	Sea Spleenwort
Fronds (30–180 cm high) rise singly from ground, not in tufts [c] on page 252	sori form faint edging to pinnules [b] on page 252	*Pteridium aquilinum*	Bracken
d, e, f, frond	Young sori oblong or sausage-shaped		
	indusium curved [d]; plant 30–100 cm high	*Athyrium filix-femina*	Lady-fern
	usu. under 12 cm, dull green; lflets fan shaped [f]	*Asplenium ruta-muraria*	Wall-rue
	10–10 cm, bright green; indusium [e]	*Asplenium adiantum-nigrum*	Black Spleenwort
	Indusium attached by its centre		
	frond rigid; about 15 lobes on longest pinna	*Polystichum aculeatum*	Hard Shield-fern
	frond soft; up to 20 lobes on a pinna	*Polystichum setiferum*	Soft Shield-fern
	Indusium ovate, pointed		
	delicate fern up to 40 cm high	*Cystopteris fragilis*	Brittle Bladder-fern

▶

Sori in neat border round pinnules		*Oreopteris limbosperma*	Lemon-scented Fern
Frond almost 3 times pinnate	stem scales w. dark centre [a] indusium [b]	*Dryopteris dilatata*	Broad Buckler-fern
	indusium w'out glands [c]	*Droyopteris carthusiana*	Narrow Buckler-fern
Pinnae w. blackish patch where they join stem	main stem densely clothed with brown scales	*Dryopteris affinis*	Scaly Male-fern
underside of frond			
Frond twice pinnate w. scales on main stem		*Dryopteris filix-mas*	Male-fern
	indusium		

Simplified key to 18 common white umbellifers

For the full key to UMBELLIFERAE and further drawings refer to page 125.

Leaves
i   once pinnate
ii twice pinnate
iii three times pinnate

flower
bracteole
ray
bract

an umbel

Radical lvs. palmate	in woods	*Sanicula europaea*	Sanicle
Lower aerial lvs. once pinnate (lflets may be deeply toothed) Ignore submerged lvs.	petals entire; bracteoles present	*Apium nodiflorum*	Fool's Water-cress
	umbels w. 3–6 rays [a]	*Sison amomum*	Stone Parsley
	usu. in water; bracts toothed; 10–15 rays [b]	*Berula erecta*	Lesser Water-parsnip
	outer fls. v. irregular; lvs. large, coarse	*Heracleum sphondylium*	Hogweed
	some lvs. twice pinnate; no bracts or bracteoles	*Pimpinella saxifraga*	Burnet-saxifrage
Stems w. purple spots	stem rough	*Chaerophyllum temulentum*	Rough Chervil
	stem smooth	*Conium maculatum*	Hemlock
Stems partly tinged w. purple	lflets broad, toothed; in fl. July–Sep.	*Angelica sylvestris*	Wild Angelica
	in fl. March–June, lvs. 2–3 times	*Anthriscus sylvestris*	Cow Parsley

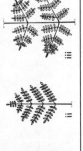

▶

Fr. bristly	bracts numerous w. long lobes [b]	*Daucus carota*	Wild Carrot
	bracts 4–12; fr. [a]	*Torilis japonica*	Upright Hedge-parsley
Bracteoles turned down, in 3's		*Aethusa cynapium*	Fool's Parsley
Lvs. once or twice 3-lobed	no bracts	*Aegopodium podagraria*	Ground-elder
Lvs. large, coarse	outer fls. v. irregular	*Heracleum sphondylium*	Hogweed
Stem smooth/striate but not grooved	6–12 rays; 2–5 bracteoles; lf. lobes narrowly linear	*Conopodium majus*	Pignut
	10–20 rays; no bracts or bracteoles	*Pimpinella saxifraga*	Burnet-saxifrage
4 or more bracts		*Oenanthe crocata*	Hemlock Water-dropwort
Fr. smooth, 4–6 mm long	4–10 rays; in fl. March–June fr. [c]	*Anthriscus sylvestris*	Cow Parsley
	12 or more rays; in fl. June–Sep. fr. [d]	*Oenanthe crocata*	Hemlock Water-dropwort
Fr. ribbed, 20–25 mm long		*Myrrhis odorata*	Sweet Cicely

fr.

a

bract

b

c

d

Simplified key to 25 common labiates (excluding the Mints)

For the full key to LABIATAE and further drawings refer to p. 166.

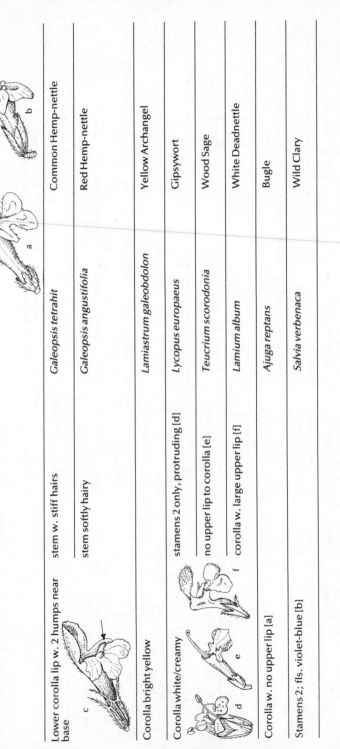

Lower corolla lip w. 2 humps near base	stem w. stiff hairs	*Galeopsis tetrahit*	Common Hemp-nettle
	stem softly hairy	*Galeopsis angustifolia*	Red Hemp-nettle
Corolla bright yellow		*Lamiastrum galeobdolon*	Yellow Archangel
Corolla white/creamy	stamens 2 only, protruding [d]	*Lycopus europaeus*	Gipsywort
	no upper lip to corolla [e]	*Teucrium scorodonia*	Wood Sage
	corolla w. large upper lip [f]	*Lamium album*	White Deadnettle
Corolla w. no upper lip [a]		*Ajuga reptans*	Bugle
Stamens 2; fls. violet-blue [b]		*Salvia verbenaca*	Wild Clary

257

▶

Calyx w. only 2 lips; fls. usu. in pairs		
fls. pink/purple; usu. under 15 cm high	*Scutellaria minor*	Lesser Skullcap
fls. blue; up to 50 cm high; calyx [a]	*Scutellaria galericulata*	Common Skullcap
At least 2 stamens longer than upper corolla lip		
calyx w. 5 almost equal teeth	*Origanum vulgare*	Marjoram
stem under fl. head hairy on 2 opposite sides	*Thymus praecox*	Wild Thyme
fl. stems hairy along edges	*Thymus pulegioides*	Large Thyme
Calyx w. 3 short and 2 much longer teeth		
corolla hooded; fls. in a terminal head	*Prunella vulgaris*	Selfheal
fls. in dense whorls w. many bracts; calyx [c]	*Clinopodium vulgare*	Wild Basil
fls. on axillary stalks, not on main stem; calyx [b]	*Calamintha sylvatica*	Common Calamint
calyx w. swollen base [d]	*Acinos arvensis*	Basil Thyme
Creeping plant w. rounded lvs.		
fls. 2–4 together	*Glechoma hederacea*	Ground-ivy
Lower corolla lip 2-lobed		
upper lvs. rounded, ½-joined in pairs	*Lamium amplexicaule*	Henbit Dead-nettle
lvs. stalked, ovate cordate fl. [e]	*Lamium purpureum*	Red Dead-nettle
Most fls. packed in 1 terminal head	*Stachys officinalis*	Betony

Fls. 12–18 mm; plant 30–80 cm high	smell strong; corolla dark purple	*Stachys sylvatica*	Hedge Woundwort
	smell strong; corolla pale purple	*Ballota nigra*	Black Horehound
	smell faint; lvs. almost stalkless	*Stachys palustris*	Marsh Woundwort
Fls. 6–8 mm	stigma longer than corolla	*Origanum vulgare*	Marjoram
	stigma inside corolla hood [a]	*Stachys arvensis*	Field Woundwort

a

Simplified key to 15 yellow compositae Dandelion type

For the full key to these yellow COMPOSITAE and further drawings refer to pages 179 and 185

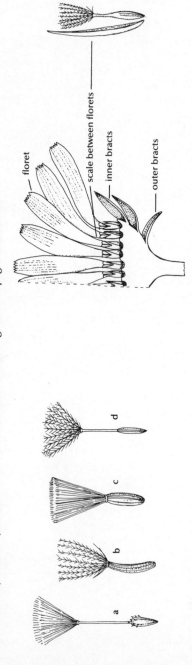

Fl. head bracts longer than florets		*Tragopogon pratensis*	Goat's-beard
Stem unbranched	stem hollow w. milky juice fr. [a]	*Taraxacum officinale*	Dandelion
	v. long white hairs on lvs. fr. [c]	*Hieracium pilosella*	Mouse-ear Hawkweed
	fl. head 12–20 cm; outer florets w'out pappus	*Leontodon taraxacoides*	Lesser Hawkbit
	fl. head 25–40 mm across fr. [b]	*Leontodon hispidus*	Greater Hawkbit
Lvs. w. pimples	fr. [d]	*Picris echioides*	Bristly Oxtongue

		Lapsana communis	Nipplewort
Fr. w'out any pappus			
No stem lvs. but several bracts below fl. head	scales among florets [a]	*Hypochaeris radicata*	Cat's-ear
	fl. head tapered below [a]; lvs. usu. narrowly lobed	*Leontodon autumnalis*	Autumn Hawkbit
Copious milky juice in fresh stems	buds v. hairy [b]; fl. head 4–5 cm across; fr. [c]	*Sonchus arvensis*	Perennial Sow-thistle
	lvs. shiny w. rounded toothed auricles [g]	*Sonchus asper*	Prickly Sow-thistle
	lvs. dark, dull w. pointed auricles [h]	*Sonchus oleraceus*	Smooth Sow-thistle
Fl. head bracts in 2 sets, 1 long 1 short [e]	in fl. March–July; fr. w. long beak [i]	*Crepis vesicaria*	Beaked Hawk's-beard
	in fl. June–Sep.; fr. not beaked [j]	*Crepis capillaris*	Smooth Hawk's-beard
Pappus pale brown [f] fr. [f]	fl. head similar to [k]	*Hieracium murorum* (agg.)	Hawkweed

Simplified key to 5 common docks

For the full key to Docks (POLYGONACEAE) and further diagrams refer to page 000.

Fr. segments distinctly toothed; teeth over 1 mm	branches spread widely; lf. blade up to 10 cm [e]; fr. [a]	*Rumex pulcher*	Fiddle Dock
	fl. clusters dense; lf. blade up to 25 cm fr. [b]	*Rumex obtusifolius*	Broad-leaved Dock
Fr. segments usu. 4 mm or more	fl. clusters dense; lvs. v. wavy edges [f]	*Rumex crispus*	Curled Dock
Fr. segments about 3 mm	fr. w. 3 warts [c]	*Rumex conglomeratus*	Clustered Dock
	fr. w. 1 wart [d]	*Rumex sanguineus*	Wood Dock

Index to plant names

The number refers to the page on which the appropriate key begins.
When a second page is given this refers to a Simplified Key.
When a plant has an English name consisting of two words both words
are indexed. Thus flowers with a two-part name may be found under such
headings as 'Water, Wood, Sea, White, Yellow, Hairy', as well as under

the second part of the name. The only exceptions to this are that the
words 'Common, Wild, Greater, Lesser, Small' are not indexed.
The order is strictly alphabetical, no notice being taken of commas,
hyphens, or spaces between words.

Index to families